D0840056

GEOPHYSICAL MONOGRAPH SERIES

David V. Fitterman, Series Editor
William H. Dragoset Jr., Volume Editor

NUMBER 7

A HANDBOOK FOR SEISMIC DATA ACQUISITION IN EXPLORATION

By Brian J. Evans

SOCIETY OF EXPLORATION GEOPHYSICISTS

Evans, Brian J., 1945 -
    A handbook for seismic data acquisition in exploration / by Brian J. Evans;
edited by William H. Dragoset, Jr.,
        p.   cm. — (Geophysical monograph series : no. 6)
    Includes bibliographical references and index.
    ISBN 1-56080-041-0 (paper : alk. paper)
    1. Seismic prospecting. I. Dragoset, William H. II. Title. III. Series.
TN269.8.E93 1997
622'.1592—dc20   96-42041
                                                                CIP

ISBN 0-931830-56-7 (Series)
ISBN 1-56080-041-0 (Volume)

Society of Exploration Geophysicists
P. O. Box 702740
Tulsa, OK 74170-2740

© 1997 by Society of Exploration Geophysicists
All rights reserved. This book or parts hereof may not be reproduced in any form
without written permission in writing from the publisher.

Published 1997
Printed in the United States of America

To my wife, Margaret, who not only typed for endless weekends to complete the earlier versions of this text but who waited at home for many months in anticipation of my returning from the field. The good thing about going away is coming home again.

Contents

Preface

In 1983, like so many other geophysical consultants of the day and the petroleum exploration industry in general, I experienced a downturn in exploration activity. In previous years, I had been an electronics engineer, a well-logging geologist/engineer, a seismic instrument engineer, a seismic crew party manager and an operations manager, after which I had turned to consulting. My "doodle-bugging" career in seismic exploration had been every young and agile person's dream—getting lost in Mexican bandit country, ducking bullets in Angola and the Philippines, playing soccer in Montevideo, cruising around Singapore in a rickshaw, visiting temples in Thailand, sitting with maidens in Senegal, dodging limpet mines in Vietnam, losing streamers in the North Sea, being incredibly inebriated in the Spratley Islands, getting stuck in the middle of the Kalahari Dessert, driving fast rental cars along interminable freeways in the United States, experiencing negative gravity during plane flights over Alberta, being caught in a force-9 gale in the Shetland Islands, losing all belongings after a "willy-willy" struck our camp in Central Australia, depositing the grand piano (and pianist) in an upmarket hotel lobby in Singapore and, in the end, having enough money to buy a fast car and a house on the same day.

I wanted to learn more about the seismic industry, which had treated me so well and which I loved so much. I had never had a formal lecture in seismic exploration, so like many others, I turned to the education industry to learn more about seismic methods so I would be better prepared when the industry picked up again. When I began studying exploration geophysics at the West Australian Institute of Technology (WAIT) in Perth, I was surprised to find there was no textbook that gave an up-to-date, in-depth treatment of seismic data acquisition—the area of geophysics in which I had spent most of my professional life. There were many good broad textbooks available, but none was an adequate lecture text. I realized quickly that I knew more about both land and marine seismic operations than was presented in most of the available texts. I had worked within and offshore from most countries of the world during the previous 11 years. The only countries I hadn't worked in were the then-communist countries. My initial duties at WAIT were to lecture (part time) in seismic data acquisition while I completed my course work. But how could I lecture without some form of text?

So, I assembled notes I had collected over the years and tried to put them in some order. I spent three months handwriting a text (there were no personal computers in those days) which I then had printed at the WAIT press. I finally started teaching seismic data acquisition to students, and I still am doing so. Meanwhile, new seismic methods and instruments were being developed, and I had to keep my notes up to date. I finally converted the handwritten text to a typed (IBM golf-ball) version, for which I am eternally grateful to my wife (since I could not type at the time). Each time I updated my version of seismic data acquisition methods and printed copies for my students, someone would prove another seismic method successful and I had to modify the text. The evolution in computing technology and its application in seismic data acquisition has been so breathtakingly fast that I haven't been able to update the text at the same rate. Therefore, dear reader, I apologize if the text still doesn't provide you with all the answers you are seeking, but I think it goes a long way toward explaining the fundamentals of our innovative science.

This book is written primarily for the novice—the person (such as me) who was qualified in another area (engineer, geologist, chemist, accountant, economist, etc.); it is pitched at students of exploration seismology who want to know how and why in simple language. I use it as my main text for final-year geophysics honors students at Curtin University of Technology. It also can be used as a good basic text for teaching seismic methods.

The text concentrates on seismic data acquisition in hydrocarbon exploration. It is light on mathematical methods but heavy on why we do things. Consequently, it will be a useful reference book for all those workers on seismic crews the world over who wonder how we possibly could get a profile through the Earth by firing shots over it. (I am still constantly amazed at what we can do with seismic.) The text does not cover refraction, shear-wave or vertical seismic profiling exploration in any detail because other books do a better job on those topics.

This book was written with Bill Dragoset's editorship, for which I am eternally grateful. I thank Western Atlas for allowing Bill time to correct a lot of my written words. I also am indebted to a few others for its publication, such as my industry colleagues from whom I have learned much, including my early field associates at Geophysical Service Inc., those at Geoservice, Aquatronics, Shell Australia, and Horizon Exploration. My current associates at Curtin University, including Norm Uren, John McDonald and Milovan Urosevic, have individually taught me a lot, as well as my colleagues from the University of Houston including Dan Ebrom, K. K. Sakharan, Bob Sheriff, and Barbara Murray, with whom I spent most of 1991.

As for me, I am a great believer in practicing what you preach. Consequently, I continue to run my experimental land crew from Curtin University, so if you ever need to talk to me on any aspect of seismic data acquisition, call me. I'm on e-mail—evans@geophy.curtin.edu.au—and I'll do my best to answer your questions. Oh yes, in return, perhaps you can update me with your latest best practice, and between us we can keep this volume updated.

Happy doodle-bugging.

Brian Evans
Senior Lecturer in Geophysics
Curtin University of Technology
Perth, Australia

Chapter 1

Seismic Exploration

1.1 Introduction

The science of seismology began with the study of naturally occurring earthquakes. Seismologists at first were motivated by the desire to understand the destructive nature of large earthquakes. They soon learned, however, that the seismic waves produced by an earthquake contained valuable information about the large-scale structure of the Earth's interior.

Today, much of our understanding of the Earth's mantle, crust, and core is based on the analysis of the seismic waves produced by earthquakes. Thus, seismology became an important branch of geophysics, the physics of the Earth.

Seismologists and geologists also discovered that similar, but much weaker, man-made seismic waves had a more practical use: They could probe the very shallow structure of the Earth to help locate its mineral, water, and hydrocarbon resources. Thus, the seismic exploration industry was born, and the seismologists working in that industry came to be called *exploration geophysicists*. Today seismic exploration encompasses more than just the search for resources. Seismic technology is used in the search for waste-disposal sites, in determining the stability of the ground under proposed industrial facilities, and even in archaeological investigations. Nevertheless, since hydrocarbon exploration is still the reason for the existence of the seismic exploration industry, the methods and terminology explained in this book are those commonly used in the oil and natural gas exploration industry.

The underlying concept of seismic exploration is simple. Man-made seismic waves are just sound waves (also called acoustic waves) with frequencies typically ranging from about 5 Hz to just over 100 Hz. (The lowest sound frequency audible to the human ear is about 30 Hz.) As these sound waves leave the seismic source and travel downward into the Earth, they encounter changes in the Earth's geological layering, which cause echoes (or reflections) to travel upward to the surface. Electromechanical transducers (geophones or hydrophones) detect the echoes arriving at the surface and convert them into electrical signals, which are then amplified, filtered, digitized, and recorded. The recorded seismic data usually undergo elaborate processing by digital computers to produce images of the earth's shallow structure. An experienced geologist or geophysicist can interpret those images to determine what type of rocks they represent and whether those rocks might contain valuable resources.

Thus, *seismic data acquisition*, the subject of this book, is just one stage of a multistage process. The full process is known as *seismic surveying*. Such surveying involves four discrete stages: survey design and planning, data acquisition, data processing, and data interpretation. The success or failure of a seismic survey often is not determined until the final interpretation stage. Because resurveying can be prohibitively expensive, it is of immense importance to ensure that all aspects of the survey are performed correctly the first time. This means that care in planning and acquiring data is extremely important. This book provides useful information for the first two stages of the survey in the hope that it will help the geophysicist to acquire the best possible data under each survey environment.

This chapter describes the fundamentals of seismic data acquisition so that the reader will gain the basic knowledge necessary to plan a sensible survey. To appreciate fully the various techniques for data acquisition, one must have a grasp of both the physics of seismic waves and the data-processing steps used to create an image of the earth. Those two topics are reviewed in this chapter. The chapter concludes with an overview of survey design considerations.

1.2   Seismic Data Acquisition

If the seismic exploration industry were to be described in one word, the word would have to be "innovative." In continually striving to improve their images of the Earth, exploration geophysicists always have been quick to adapt new technologies in electronics, computer processing, data recording, and transducer design to seismic surveying. Because of this innovative spirit, seismic exploration technology has evolved rapidly, especially during the past 30 years. Today a well-rounded exploration geophysicist should have a basic knowledge of not only geology but also of physics, mathematics, electronics, and computer science. Grasping mathematics at a high level was not required of the earliest seismologists.

1.2.1 Historical Perspective

The earliest known seismic instrument, called the seismoscope, was produced in China about A.D. 100. A small ball bearing was wedged in the mouth of each of six dragons mounted on the exterior of a vase (Figure 1). An earthquake motion would cause a pendulum fastened to the base of the vase to swing. The pendulum in turn would knock a ball from a dragon's mouth into a toad's mouth to indicate the direction from which the tremor came.

In 1848 in France, Mallet began studying the Earth's crust by using acoustic waves. This science developed into what is now called earthquake seismology, solid earth, or crustal geophysics, which is still a broad area of academic research. In 1914 in Germany, Mintrop devised the first seismograph; it was used for locating enemy artillery during World War I. In 1917 in the United States, Fessenden patented a method and apparatus for locating ore bodies.

The introduction of refraction methods for locating salt domes in the Gulf Coast region of the United States began in 1920, and by 1923, a German seismic service company known as Seismos went international (to Mexico and Texas) using the refraction method to locate oil traps.

As the search for oil moved to deeper targets, the technique of using reflected seismic waves, known as the *seismic reflection method*, became more popular because it aided delineation of other structural features apart from simple salt domes. It is said that one of the few good things produced during World War II was the technological advances made in seismology in the search for oil. Because Germany could not overcome the Allied forces holding Middle East oil supplies, it had to develop indigenous oil to sustain the war effort. As a result, the Seismos company's budget was increased. The results

Fig. 1. The seismoscope (after Sheriff).

included several technical innovations that furthered the development of seismic data acquisition equipment and the interpretation of seismic data.

Beginning in the early 1930s seismic exploration activity in the United States surged for 20 years as related technology was being developed and refined (Figure 2). For the next 20 years, seismic activity, as measured by the U.S. crew count, declined. During this period, however, the so-called digital revolution ushered in what some historians now are calling the Information Age. This had a tremendous impact on the seismic exploration industry. The ability to record digitized seismic data on magnetic tape, then process that data in a computer, not only greatly improved the productivity of seismic crews but also greatly improved the fidelity with which the processed data imaged earth structure. Modern seismic data acquisition as we know it could not have evolved without the digital computer.

During the past 20 years, the degree of seismic exploration activity has become related to the price of a barrel of oil, both in the United States (Figure 3) and worldwide. In 1990, US$2.195 billion was spent worldwide in geophysical exploration activity (Goodfellow, 1991). More than 96% of this (US$2.110 billion) was spent on petroleum exploration.

Despite the recent decline in the seismic crew count, innovation has continued. The late 1970s saw the development of the 3-D seismic survey, in which the data imaged not just a vertical cross-section of earth but an entire volume of earth. The technology improved during the 1980s, leading to more

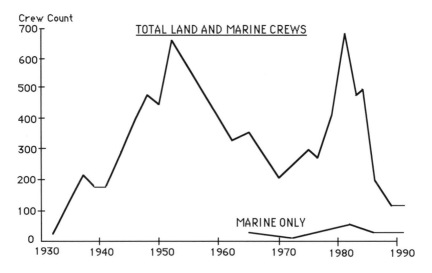

Fig. 2. U.S. seismic crew count (Goodfellow, 1991).

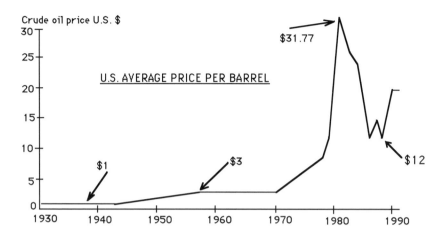

Fig. 3. U.S. price per barrel (courtesy U.S. Bureau of Mines, API).

accurate and realistic imaging of earth. This was partly responsible for the increased use of seismic data by the production arm of the oil industry.

1.2.2 Modern Data Acquisition

Because subsurface geologic structures containing hydrocarbons are found beneath either land or sea, there is a land data-acquisition method and a marine data-acquisition method. The two methods have a common goal—imaging the earth. But because the environments differ so, each requires unique technology and terminology.

In this section, simple examples of both methods are described in a presentation of the basic concepts of seismic data acquisition. Also, a hybrid of the two methods, called *transition-zone recording,* is described briefly.

Consider the simple land acquisition diagram shown in Figure 4. A seismic wave is generated by exploding an energy source near the surface to cause a shock wave to pass downward toward the underlying rock strata. Some of the shock wave's energy is reflected from the rocks back to the surface. The geophones vibrate as the reflected seismic wave arrives, and each generates an electrical signal. This signal is passed along cables to a recording truck, where it is digitized and recorded on magnetic tape or disk. The recorded information is taken to a computer center for processing. The seismic recording technique often is referred to as seismic surveying, so the words "recording" and "surveying" are interchangeable.

The positions at which the energy sources are detonated are called *shotpoints.* The energy-receiving geophones—"phones" for short—are placed

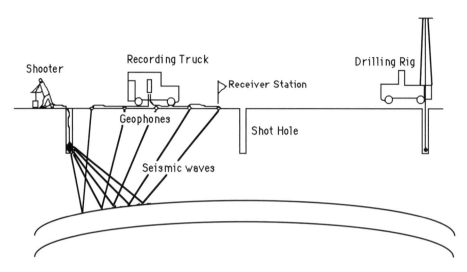

Fig. 4. Seismic land survey using an explosive source.

along a line at points known as *stations*. The geophones are electrically connected to the recording truck (known as a "doghouse" or "dog-box") by
cables. The recording truck engineer is referred to as the *observer*, and he and a
line foreman organize the placement and retrieval of the geophones by personnel called "juggies.

Each station's location must be known, so a surveyor and assistants are
used to survey the line prior to recording. The *survey party* places wooden
pegs at the stations along this line. These pegs define the location of the seismic line or seismic section to be shot. Sometimes a drilling crew is required if
shot holes are needed. A *party manager* controls daily operations, and line-
kilometers or line-miles of seismic profile are recorded daily by the seismic
party or crew.

A display of the received data is called a *shot record* and usually consists of
wiggle traces, where each trace represents the electrical output signal from a
geophone. Figure 5 shows a land shot record that has 96 such traces, with 48
of them being on each side of the energy source. Such records are displayed as
they are recorded over a period of time, so while the horizontal axis of the
record shows the traces from 96 stations, the vertical axis is time. Reflected
energy arrives in waves; some are labeled on the figure. The display is produced by a seismic *camera,* so called because in the early years of the seismic
technique, traces were recorded onto film that was developed in a darkroom
and hung out to dry like normal photographs.

In this record, a seismic reflection event is seen at A around 600 ms at the
traces on the left; this event appears at about 450 ms near the center of the

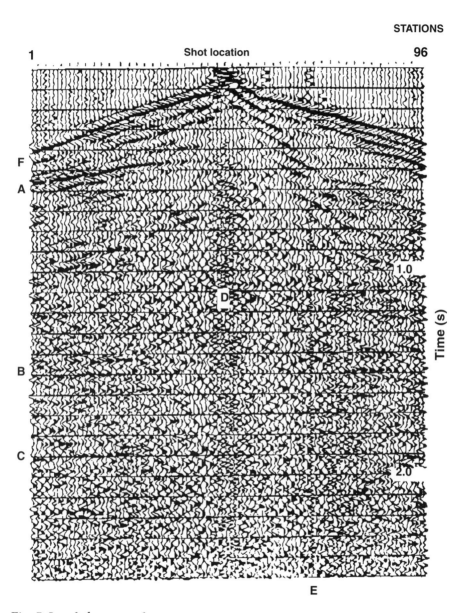

Fig. 5. Land shot record.

record. So, the reflected energy has arrived as a seismic wave that has passed across the stations to the left of the shotpoint. Deeper reflection events are observed to be arriving at B near 1450 ms and at C near 1900 ms on the left half. There are also a number of events on the right half, although they do not stand out as clearly as those on the left. Several traces near the shot, at D, and also one trace at E, are especially noisy. The event at F is the first energy arrival. That energy may be the result of a seismic wave traveling horizontally from the source to the phones (a *direct arrival*) or energy that has refracted along a shallow layer boundary in the earth (a *refraction arrival*).

Using the information in Figure 5, the velocity at which the direct-arriving wave traveled may be computed. The distance from the shotpoint to any geophone is called that phone's *offset* distance (hence, the distance from the shotpoint to the nearest phone is the near-offset distance, while the distance from the shotpoint to the farthest phone is the far-offset distance). In Figure 5, therefore, the near-offset distance is the distance from the shotpoint location, midway between stations 48 and 49, to those stations (i.e., a half-station distance); the far-offset distance is the distance from the shotpoint midway between 48 and 49 to the farthest stations at 1 and 96 (i.e., 47.5 station lengths). The time taken for the direct arrival F to travel from the source to the far station 1 is 400 ms. To compute the arrival velocity for F, the distance traveled is divided by the time of travel. Therefore, if the distance between stations (called the *station interval*) is 12.5 m, then the velocity of event F is (47.5 x 12.5 m) / 0.4 s = 1484.4 m/s. This happens to be close to the velocity of sound through water, so we may assume that the direct wave probably traveled along a water table (or through water-saturated soil) situated just beneath the surface of the earth.

Thus, a shot record not only shows the presence (or in some cases the absence) of various kinds of seismic events but also allows for determination of the propagation velocity through the earth. The propagation velocity (or speed of sound) in a rock layer is indicative of the type of rock. For linear events such as F in Figure 5, the velocity is given simply by dividing the distance traveled by the time of travel. Reflection events such as A have a more complicated relationship among velocity, time, and offset because their travel paths (Figure 4) are not a straight line. As will be seen later, though, analysis of reflection events also yields information about the propagation velocity of sound in the earth.

As with land acquisition, marine recording is performed by exploding an energy source and recording reflected energy. Because the recording process takes place offshore, a ship tows the energy source and phones behind it (Figure 6). All members of the seismic crew are aboard the ship. The cable towed astern is a *streamer* containing the hydrophones. The ship's position is typically monitored by radio navigation so that shots (or "pops") can be fired at the desired locations.

Just as with land records, marine shot records also are recorded and displayed in time (Figure 7). Instead of traces showing stations versus time, they are referred to as *channels* versus time. The shot records in Figure 7 have the ship and energy-source position to the left of the streamer. Seismic events such as A arrive first at channels on the left which are nearest to the source, then spread to the right in a curved manner. Event B is the direct arrival. The area of a marine shot record of greatest interest to the geophysicist is windowed on the right-hand record. A comparison of the land shot record (Figure 5) with the marine records shows that the marine events appear more continuous across the record. Although some reflection events are visible on the land record, most of that record is obscured by surface-generated noise. The marine record—being relatively noise free—is said to have a high signal-to-noise ratio, while the land record has a low signal-to-noise ratio. Reasons for this are discussed in greater detail in Chapter 3.

Consider again the land and marine acquisition schemes (Figures 4 and 6). After each land shot, the line of receivers may be moved along to another appropriate location and the shot fired again. This is the so-called *roll-along method* of seismic recording, the parameters of the roll-along being governed by both the geology and how the data are to be processed. Alternatively, the geophones may be left in place while the shot position is moved several times. To record an extensive number of lines on land is clearly time consuming because of the need to reposition the geophones manually. In marine

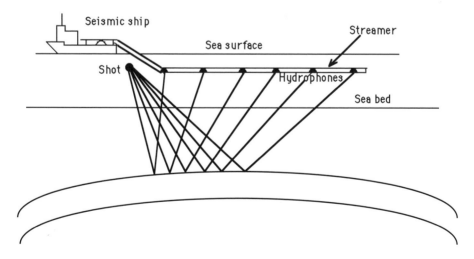

Fig. 6. Marine recording technique.

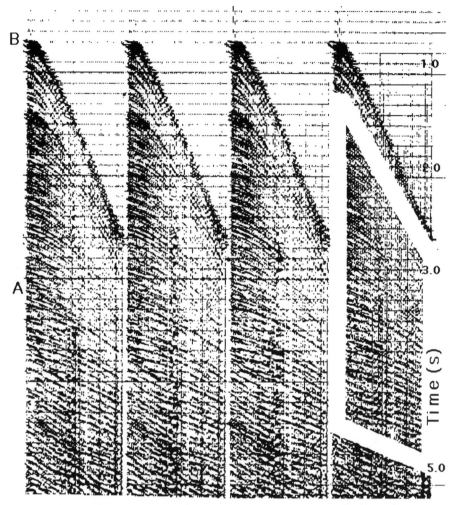

Fig. 7. Marine shot records. (Courtesy of Allied Geophysical Laboratories, University of Houston.

acquisition, however, both the sources and receivers are very mobile. Typically, only 10 seconds lapse between one shot and the next. Thus, on a per-kilometer basis, marine acquisition is much less costly than land acquisition. On the other hand, start-up costs for a marine crew are higher than those for a land crew because of the cost of the ship.

The operational difficulties faced by land and marine crews differ considerably. Marine crews are beset with the problems of keeping complex equipment performing well in the harsh ocean environment; land crews are more likely to be hindered by cultural hazards such as the need to avoid disrupting a region's agricultural activities. For these reasons, land and marine operations often are considered as separate endeavors and the field personnel involved in one type of operation rarely move into the other. The case in which both land and marine operations are conducted together is known as transition-zone recording. Transition-zone recording takes place in the coast-line area where the land line is terminated by the sea and shallow sea depths restrict access by a standard marine seismic vessel. Following is a more detailed contrast of the three survey types.

1.2.2.1   Land Data Acquisition

In land acquisition, a shot is fired (i.e., energy is transmitted) and reflections are recorded at a number of fixed receiver stations. These geophone stations are usually *in-line* although the shot source may not be. When the source is in-line with the receivers—at either end of the receiver line or positioned in the middle of the receiver line—a two-dimensional (2-D) profile through the earth is produced. If the source moves around the receiver line causing reflections to be received from points out of the plane of the in-line profile, then a three-dimensional (3-D) image is possible (the third dimension being distance, orthogonal to the in-line receiver line). Land operations are relatively slow compared with the 24-hour-per-day recording that takes place in marine seismic surveying. The majority of land survey effort is expended in moving the line equipment along across farm fields or through populated communities. Hence, operations often are conducted only during daylight.

1.2.2.2   Marine Data Acquisition

In a marine operation, a ship tows one or more energy sources astern parallel with one or more towed seismic receiver lines. In this case, the receiver lines take the form of cables containing a number of hydrophones. The vessel moves along and fires a shot, with reflections received by the streamers. If a single streamer and a single source are used, a single seismic profile may be recorded in like manner to the land operation. If a number of parallel sources and/or streamers are towed at the same time, the result is a number of parallel lines recorded at the same time. If many closely spaced parallel lines are recorded, a 3-D volume of data is recorded. More than one vessel may be employed to acquire data. Marine operations usually are conducted on a 24-hour basis since there is no need to curtail operations in the dark.

1.2.2.3   Transition-Zone Recording

Because ships are limited by the water depth in which they safely can conduct operations, and because land operations must terminate when the source approaches the water's edge, transition-zone recording techniques must be employed if a continuous seismic profile is required over the land and then into the sea. Geophones that can be placed on the seabed are used with both marine and land shots fired into them. As may be imagined, different types of coastline require different equipment; consequently, these operations are often more labor intensive than either land or marine operations. They also can be the most expensive to record and process because of operational and instrumentational complexities.

In transition-zone surveys, any number of shallow draft vessels are employed and operations usually are conducted 12 hours daily. This book will concentrate on describing and contrasting land and marine operations because, in most cases, transition-zone surveying is merely a mixture of the two.

Although the 1989 marine crew count (Figure 2) appears low compared with the total number of seismic crews, marine crews recorded three times more seismic data than land crews, as shown in Table 1:

Table 1.1. 1989 crew count statistics.

Crew type	Miles recorded	Avg. cost per mile (US$)
Land	241 265	3511
Marine	777 278	700
Transition zone	9700	1956

1.3 Seismic Wave Fundamentals

Before seismic surveying methods are discussed further, the reader should gain a basic understanding of seismic waves themselves. This section will equip the reader with a fundamental knowledge of the phenomena of wave propagation. It reviews the various seismic wave types, describes how they propagate and are affected by changes in geology, and discusses the ray-tracing concept. Energy-decay considerations then are reviewed. The seismic phenomena of reflection, refraction, transmission, and diffraction are discussed. The section finishes with an explanation of the problem of seismic multiples and a description of the so-called *normal moveout* (NMO) of events in a seismic shot record.

1.3.1 Types of Elastic Waves

To understand the phenomena of seismic wave transmission, consider an explosive detonating in a shot hole (Figures 8 and 9). After the initial fracturing of the hole around the exploding energy point, further transmission of energy can be explained by assuming the Earth has the elastic properties of a solid. The Earth's crust is considered as completely elastic (except in the

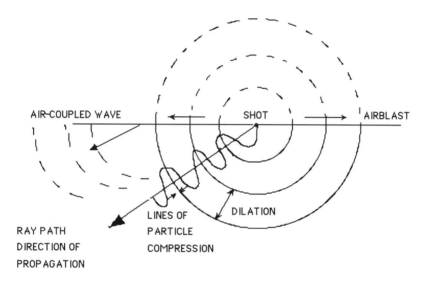

Fig. 8. Compressional wave transmission.

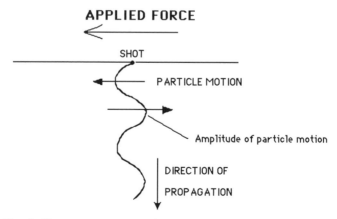

Fig. 9. Shear wave generation.

immediate vicinity of the shot), and hence the name given to this type of acoustic wave transmission is *elastic wave propagation*. Several kinds of wave phenomena can occur in an elastic solid. They are classified according to how the particles that make up the solid move as the wave travels through the material.

1.3.1.1   Compressional Waves (*P*-waves)

On firing an energy source, a compressional force causes an initial volume decrease of the medium upon which the force acts. The elastic character of rock then causes an immediate rebound or expansion, followed by a dilation force. This response of the medium constitutes a primary *compressional wave* or *P*-wave. If we were able to put a finger against the rock in line with the *P*-wave arrival, our finger would move back and forth in the direction of wave propagation, just like the particles that make up that rock. Particle motion in a *P*-wave is in the direction of wave propagation. The *P*-wave velocity is a function of the rigidity and density of the medium. In dense rock, it can range from 2500 to 7000 m/s, while in spongy sand, from 300 to 500 m/s.

In addition, on land the energy source (shot) generates an airwave known as the *air blast*, which itself can set up an *air-coupled wave*, a secondary wavefront in the surface layer. This wave generally travels at about 350 m/s, a slower velocity than the compressional wave. The speed of the airwave, which depends mainly on temperature and humidity, varies from 300 to 400 m/s.

1.3.1.2   Shear Waves (*S*-waves)

Shear strain occurs when a sideways force is exerted on a medium; a *shear wave* may be generated that travels perpendicularly to the direction of the applied force. Particle motion of a shear wave is at right angles to the direction of propagation. A shear wave's velocity is a function of the resistance to shear stress of the material through which the wave is traveling and is often approximately half of the material's compressional wave velocity.

In liquids such as water, there is no shear wave possible because shear stress and strain cannot occur. Marine records generally appear to have higher signal-to-noise ratios than land records. This is partly because in marine recording, since shear waves are not generated in the water by the source or received by the hydrophones, all arrivals are compressional waves. Shear waves are readily generated and received during land operations; land records often contain a mixture of compressional and shear waves as well as other types of waves.

With stratified rock in which there are fluid-filled cracks and inclusions, there is often a greater resistance to a shear force than in a homogenous rock, as cracks limit the degree of shear particle movement. The result is that a shear velocity change may occur as a result of the layering and cracking. Compressional waves are not so readily affected by cracks. A comparison of compressional velocities with shear-wave velocities in such media therefore conveys information about the nature of the rock. Obtaining such information is the goal of the type of seismic surveying called *shear-wave exploration*.

1.3.1.3 Mode-Converted Waves

Each time a wave arrives at a boundary, a portion of the wave is reflected and transmitted. Depending upon the elastic properties of the boundary, the *P*-wave or *S*-wave may convert to one or the other or to a proportion of each. Such converted waves sometimes degrade the signal-to-noise ratio. This degradation causes problems during data processing.

1.3.1.4 Surface Waves

On land, the weathering of surface rocks and the laying down of soft sediment over the years causes a layer of semiconsolidated surface rock overlying the sedimentary section to be explored. This layer is known as the *weathering layer* or *low-velocity layer* (LVL). The latter term is used because of the low velocity of propagation of *P*-waves passing through the layer. The LVL also allows the transmission of surface waves along its air-earth boundary. Surface waves spread out from a disturbance like ripples seen when a stone is dropped into a pond.

Lord Rayleigh (1842-1919) developed the physics to explain surface waves; in his honor the surface wave is now commonly known as the *Rayleigh wave*. Dobrin (1951) performed a series of trials to test Rayleigh's theory. He suspended geophones down vertical boreholes and fired a number of explosive shots at the surface. His measurements of the relative amplitude and direction of particle motion agreed with the theory that Rayleigh waves were of low frequency, traveling horizontally with retrogressive elliptical motion and away from the energy source (shot), as shown in Figure 10.

Going deeper (i.e., down the bore), Dobrin found that the particle motion of the surface wave reduced in amplitude with increases in depth, eventually reversing in direction. This point was in the vicinity of the base of the weathering layer. Because the motion of the ground appears to roll, the wave is commonly known as *ground roll*.

Figure 11 indicates the Rayleigh wave's elliptical ground-roll motion, and Figure 12 shows how surface waves appear on a shot record; several surface-wave modes can be seen (a and b). The clearly defined reflections at c and d are completely masked by the surface waves at the shorter offsets.

SHOT

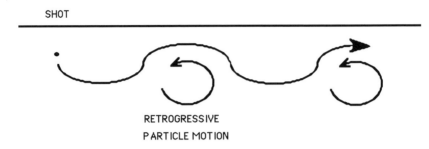

RETROGRESSIVE
PARTICLE MOTION

Fig. 10. Surface wave motion.

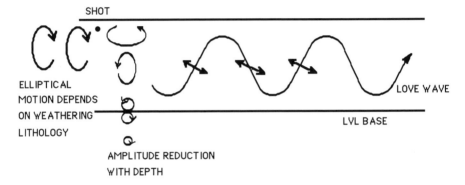

Fig. 11. Weathering layer wave motion.

Such surface waves appear as coherent events on seismic reflection records, where they are treated as unwanted noise. In some regions where the weathering layer is thick, ground roll completely masks useful reflected data. In such areas, the signal-to-noise ratio is therefore very poor and the resultant seismic section is often equally poor in defining a sequence of seismic reflections. Offshore surveys often observe Rayleigh wave equivalents (Scholte waves) as long-period, water-bottom, sinusoidal waves, known as *bottom roll* or *mud roll*. This tends to occur only in water depths of 10–20 m.

1.3.1.5   Love or Pseudo-Rayleigh Waves

The *Love wave* is a surface wave borne within the LVL, which has horizontal motion perpendicular to the direction of propagation with, theoretically, no vertical motion. Also known as the horizontal *SH*-wave, *Q*-wave, *Lq*-wave, or *G*-wave in crustal studies, such waves often propagate by multiple reflection within the LVL, dependent upon the LVL material (Figure 11). If such

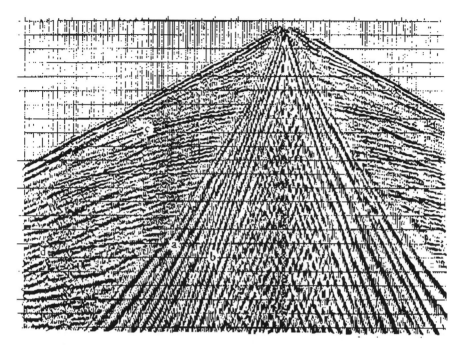

Fig. 12. Shot record showing extreme ground roll obscuring reflection events.

waves undergo mode conversion, a number of noise trains appear across the seismic record, obscuring reflected energy content even further.

1.3.1.6 Direct and Head Waves

The expanding energy wavefront that moves along the air-surface interface outward from a shot commonly is observed as the *direct wave* and has the velocity of the surface layer through which it travels. In the marine case, the direct wave has been used to determine the speed of sound in water, which is around 1500 m/s. *Head waves* are the portions of the initial wavefront that are transmitted down to the base of the weathering layer or the water bottom and are refracted along the weathering base. They then return to the surface as refracted energy or *refractions*. Sometimes the refracted velocity is higher than the velocity of propagation in the surface layer. In that case, refracted head waves appear in the mid- to far-offset traces before arrival of the direct wave.

1.3.1.7 Guided Waves

When a layer of the Earth has an extreme density or velocity contrast at both its upper and lower boundaries, a wave traveling along the layer may undergo internal reflection (i.e., stay within the layer, reflecting from upper interface to lower, back up again, and so on). Such waves are called *guided waves* and exhibit mainly vertical particle motion. They appear as short shingled waves, repeating on the shot record.

1.3.1.8   Ray Theory

Seismic waves created by an explosive source emanate outward from the shotpoint in a 3-D sense. If the spherically expanding 3-D wave is to be understood, a simple mechanism is needed to explain how the wave responds on contact with geological discontinuities. Huygens's principle is commonly used to explain the response of the wave when a mathematically rigorous treatment is not necessary. Every point on an expanding wavefront can be considered as the source point of a secondary wavefront (Figure 13). The envelope of the secondary wavefronts produces the primary wavefront after a small time increment. The trajectory of a point moving outward is known in optics as a *ray*, and hence in seismics as a *raypath* (Figure 14). Thinking of waves in terms of raypaths allows the geophysicist to model the wave numerically and visually more simply than if the full wavefront were considered.

The mathematical explanation of the 3-D elastic wave may be found in standard textbooks, for example, Aki and Richards (1980), Pilant (1979), Kennett (1983), Bullen and Bolt (1987). The theoretical treatment for raypath modeling is given by Cerveny et al. (1977). Such modeling is also known as *ray-trace modeling.*

Hereafter, the raypath concept is used to explain what happens as a wavefront expands. For example, the wavefront energy gradually decays in amplitude as it passes through rock. The next section will discuss this amplitude decay.

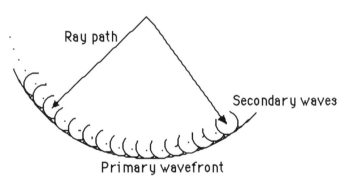

Fig. 13. Huygens's principle.

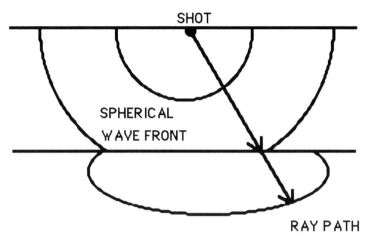

Fig. 14. Raypath trajectory.

1.3.1.9 Energy Decay

There are two types of amplitude decay, namely *spreading loss* and *absorption*. As a seismic wave expands outward from a shot, the energy per unit area of the wavefront is inversely proportional to the square of the distance from the shot because the total energy—which remains constant if absorption is ignored—has to spread over an increasingly larger area (Figure 15). This phenomenon is called energy spreading loss. The amplitude of a wave is proportional to the square root of the energy per unit area. Thus, a wave's amplitude is inversely proportional to the distance traveled.

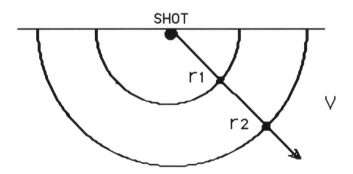

Fig. 15. Spreading loss.

As a wavefront expands, the spreading loss per unit area is given by

$$\text{Amplitude Spreading Loss in dB} = 10\log\frac{r_2}{r_1}, \tag{1}$$

where r_1 and r_2 are two radial distances from a shot (see Appendix A for an explanation of decibel). A correction to recover this amplitude loss is applied in data processing. Such processing algorithms are based on equation (1).

Amplitude loss also occurs as a wavefront passes through the rock, vibrating the rock particles. The vibrating particles absorb energy as heat; this form of energy loss is called absorption. Amplitude loss because of absorption varies exponentially with distance, so that in amplitude terms

$$A_2 = A_1 e^{-\alpha(r_2 - r_1)}. \tag{2}$$

where A_1 is the amplitude at distance r_1, A_2 is the amplitude at distance r_2, and α is the absorption coefficient of the material.

$$\text{Absorption loss in dB} = 10\log e^{-\alpha(r_2 - r_1)} = 4.3\alpha(r_2 - r_1). \tag{3}$$

Accounting for both types of losses, we have in amplitude terms

$$A_2 = A_1 \frac{r_1}{r_2} e^{-\alpha(r_2 - r_1)}. \tag{4}$$

Assuming a typical value of α is 0.25 dB per wavelength λ, where r_1 and r_2 are the radial distances from the shot, V is the velocity of sound through the material, f is the wavefront's predominant frequency, and wavelength $\lambda = V/f$, then from equation (3),

$$\text{Absorption loss in dB} = \frac{1.1f(r_2 - r_1)}{V}. \tag{5}$$

Thus, for a fixed velocity, absorption loss is frequency and distance dependent. For a particular frequency and velocity, absorption loss may be greater or less than the spreading loss. At short radial distances from a shot (shallow depth), the spreading loss is greater than the absorption loss because the logarithmic ratio in equation (1) is greater than the linear ratio in equation (5). At greater depth, the absorption loss tends to be greater than the spreading loss.

According to equation (5), higher frequencies experience greater absorption loss than do lower frequencies. The combined effect of both losses explains why deeper events in shot records generally have lower frequency content.

1.3.1.10 Reflections

Energy incident on a subsurface discontinuity is both transmitted and reflected. The amplitude and polarity of reflections depend on the acoustic properties of the material on both sides of the discontinuity (Figure 16). Consider the boundary between two layers of sonic velocities V_1 and V_2, and densities ρ_1 and ρ_2.

Acoustic impedance is the product of density and velocity. The relationship among incident amplitude A_i, reflected amplitude A_r, and reflection coefficient R_c is:

$$A_r = R_c A_i,\qquad(6)$$

where

$$R_c = \frac{\rho_2 V_2 - \rho_1 V_1}{\rho_2 V_2 + \rho_1 V_1}.\qquad(7)$$

Equation (7) shows that R_c ranges from –1 to +1 and is negative when the second layer has lower acoustic impedance than the first layer. When a reflection coefficient is negative, the polarity of the reflected wave is reversed from

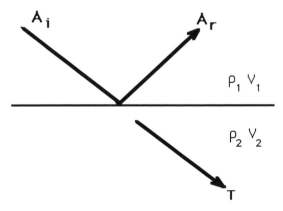

Fig. 16. Ray at a boundary.

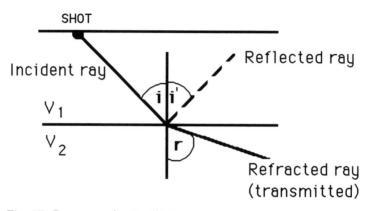

Fig. 17. Geometry for Snell's Law.

that of the incident wave. Where velocity is constant, a density contrast will cause a reflection, and vice versa. In other words, any abrupt change in acoustic impedance causes a reflection to occur. Energy not reflected is transmitted. With a large R_c, less transmission occurs and, hence, signal-to-noise ratio reduces below such an interface. Ideally, geophysicists would prefer that the earth layering had small R_c's increasing in size gradually with depth so as to compensate for the spreading and absorption losses with increased depth. Exercise 1 at the end of this chapter provides the reader with some practice in using equation (7).

When an impinging wave arrives at an interface, part of its energy is reflected back into the same medium. The incident angle i is then equal to the reflection angle i' as shown in Figure 17.

1.3.1.11   Snell's Law

Snell's Law describes how waves refract. Simply put, Snell's Law states that the sine of the incident angle of a ray, sin i, divided by the initial medium velocity V_1 equals the sine of the refracted angle of a ray, sin r, divided by the lower medium velocity V_2 (Figure 17). That is,

$$\sin i / V_1 = \sin r / V_2. \tag{8}$$

Thus, when a wave encounters an abrupt change in elastic properties, part of the energy is reflected, and part of the energy is transmitted or refracted with a change in the direction of propagation occurring at the interface.

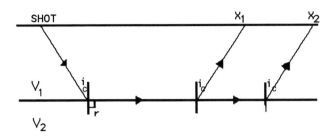

Fig. 18. Refraction raypaths.

1.3.1.12 Refractions

When an impinging wave arrives at such an angle of incidence that energy travels horizontally along the interface at the velocity of the second medium, then *critical reflection* occurs. The incident angle, i_c, at which critical reflection occurs can be found using Snell's Law:

$$\sin i_c = V_1/V_2 \sin 90° = V_1/V_2. \tag{9}$$

Figure 18 shows a typical raypath when angle i equals or exceeds i_c. The arrivals at x_1, x_2, etc. are refractions. Refraction data are useful for determining the LVL depth, dip, and velocities. However, where complex LVL bodies exist, the refraction method for the determination of weathering information loses accuracy. In such cases, deep uphole surveys often are preferred.

1.3.1.13 Diffractions

Diffractions occur at sharp discontinuities, such as at the edge of a bed, fault, or geologic pillow. Consider a wavefront arrival at a point discontinuity as shown in Figure 19. When the wavefront arrives at the edge, a portion of the energy travels through into the higher velocity region, but much of it is reflected, as shown. The reflected wavefront arrives at the receiving elements to give a curved event on the seismic record as shown in Figure 20.

In conventional in-line recording, diffractions may arrive from out of the plane of the seismic line's profile. Such diffractions are considered noise and reduce the in-line signal-to-noise ratio. However, in 3-D recording, in which specialized data processing techniques are used (i.e., 3-D seismic migration), the diffractions are considered as useful scattered energy because the data-processing routines transfer the diffracted energy back to the point from which it came, thereby enhancing the subsurface image. Hence, in 3-D surveying, out-of-the-plane diffracted events are considered part of the signal.

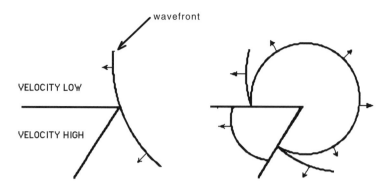

Fig. 19. Wave arrival and diffraction transmission.

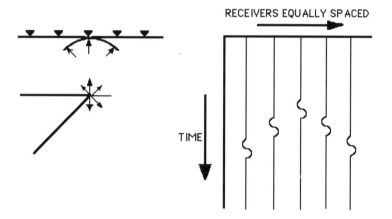

Fig. 20. Point diffraction and its seismic record.

1.3.1.14   Multiple Raypaths

Offshore seismic surveys often are affected by air-sea surface/water-bottom multiples and interbed (or "peg-leg") multiples. The seabed multiples are caused by high acoustic impedance contrasts between the air and sea, and between the sea and seabed, that reflect much of the incident energy. In particular, the air-sea interface is completely reflective with a reflection coefficient R_c of -1 (dimensionless units). This means a $180°$ phase change of the wavefront occurs at the surface (i.e., polarity reversal). A seabed reflection coefficient of 0.5 is not uncommon and is high compared to the normal R_c range of 0.01 to 0.1 found in the subsurface. Interbed multiples can occur when a portion of the incident wavefront energy becomes internally reflected within a layer.

Figure 21 shows both water-bottom multiples and interbed multiples. The first water-bottom multiple is reversed polarity because of its reflection off the sea surface. The second water-bottom multiple undergoes another reversal, which returns its polarity to that of the original wavefront. The interbed multiple may or may not undergo such a phase reversal depending upon the sign of the reflection coefficients at the top and bottom of the bed.

The time taken for an initial water-bottom reflection to be received at the streamer is the two-way traveltime for the expanding wavefront to travel from the energy source down to the seabed and back up to the streamer. The first water-bottom multiple has twice the zero-offset traveltime of the initial water-bottom reflection. The nth water-bottom multiple has a traveltime equal to $n + 1$ times that of the water-bottom reflection.

Often, distinguishing between multiples and true reflections is not easy. The problem with multiples is that they are an unwanted noise that have the same appearance as normal reflections and have similar traveltimes. For example, in the marine shot record in Figure 7, the event at 700 ms is the sea-bed reflection, while the events at about 1.4 s and 2.1 s are probably seabed multiples.

Ideally, such multiples are best removed in the field if possible. By careful choice of streamer and source depth, multiples sometimes can be attenuated. Alternatively, the source output power or streamer group spacing may be

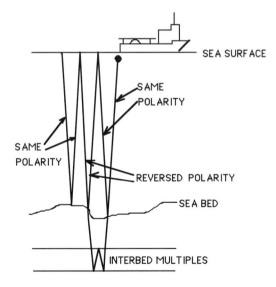

Fig. 21. Multiple raypaths for a positive seabed reflection coefficient.

adjusted to reduce these effects. Whichever technique is used, field testing becomes a requirement. Such testing requires the expertise of experienced geophysicists aboard the vessel and, of course, it costs time and money. More often than not, multiple noise cannot be suppressed adequately by field techniques. In that case, the multiple noise becomes a problem at the data-processing stage of a seismic survey. There are several processing methods that can remove multiples. One example is the NMO correction, followed by stacking.

1.3.1.15 Normal Moveout (NMO)

Reflected events on a shot record do not appear as straight lines but as curved lines (see Figure 7). This effect is called normal moveout (NMO). As shown in Figure 22, because reflected wave b has a longer travel path than reflected wave a, a horizontal bed appears on a record as a hyperbolic curve. The shape of the curve is a function of the velocity of sound in the Earth and the depth and dip of the reflecting interface.

Consider the geometry of Figure 22, which shows two rays from the shot. Wave a travels down to point d and reflects up to arrive at a phone positioned a distance x away from the shot. Suppose two imaginary lines are constructed on the left part of Figure 22. One extends the line from point x to d below the reflecting interface, and the other extends vertically below the shotpoint. They meet at the so-called image point, I. The distance from the shot to d is the same as the distance from the image point I to d. Hence, the distance traveled by wave a from the shotpoint to d and up to the phone is the same as the distance from the image point I to the phone. Also, distance h (the depth to the reflecting layer) is the same as the distance from that reflector to I. Hence, the distance from shot to image point is $2h$.

By the construction of the image point lines, a Pythagorean triangle has been drawn. In that triangle, of sides $2h$ vertically and x horizontally, the hypotenuse D may be computed

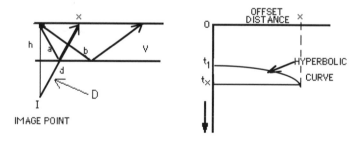

Fig. 22. Normal moveout.

$$D = \sqrt{(2h)^2 + x^2}, \tag{10}$$

or

$$D = \sqrt{4h^2 + x^2}. \tag{11}$$

If the reflection was received by a phone positioned at the shotpoint ($x = 0$), the traveltime would be that for zero offset. This is known as the *zero-offset time*. In Figure 22, the zero-offset reflection follows a vertical raypath from the source to the reflecting layer and back up the same path to the receiver at the shotpoint location. The synthetic shot record of Figure 22 shows this event as occurring at time t_1. Hence,

$$t_1 = \frac{2h}{V}, \tag{12}$$

or

$$h = \frac{t_1 V}{2}, \tag{13}$$

where V is the average speed of sound in the earth.

Substituting this expression for h in equation (11) yields

$$D = \sqrt{t_1^2 V^2 + x^2} \tag{14}$$

or

$$\frac{D}{V} = \sqrt{t_1^2 + \frac{x^2}{V^2}}. \tag{15}$$

Defining t_x as the traveltime for an event received from the reflecting layer for a phone at offset x, gives

$$t_x = \sqrt{t_1^2 + \frac{x^2}{V^2}}, \tag{16}$$

or

$$t_x^2 = t_1^2 + \frac{x^2}{V^2}. \tag{17}$$

Equation (17) describes a hyperbolic shape geophysicists call the normal moveout hyperbola, or simply NMO. It describes the relationship of the arrival time of a reflection event to the reflector's depth (via t_1), the source-to-phone offset, and the average speed of sound in the earth layers through which the wavefront travels. Because t_x represents the total traveltime for a reflection—that is, the sum of time the wave travels downward and the time it travels upward—t_x is called the *two-way traveltime.*

During data processing NMO is removed from the data by shifting each trace sample upward by an amount $\delta_t = t_x - t_1$. The quantity δ_t is called the *NMO correction.* Further discussion of NMO appears in Section 1.4.

1.3.1.16   Events on a Shot Record

The various types of seismic events that are observed on a shot record are summarized in Figure 23. Note that the only useful event—the primary reflection—must compete with all of the other wave types so far discussed. These other wave types commonly are referred to as *coherent noise.* Other forms of noise also are prevalent in day-to-day seismic recording; there are various ways to try to attenuate such noise, as listed in Figure 23.

Direct arrivals and ground-roll travel from the shotpoint horizontally. Thus, their arrival times across a receiver spread represent *one-way time* rather than the two-way traveltime associated with reflection events.

1.4   The Common Midpoint Method

The *common midpoint method* of seismic surveying is universally accepted as the optimum approach to obtaining an image of earth layers. When a shot is fired, the emanating wave has many rays that travel downward. When the incident wave is reflected from a horizontal boundary, the point of reflection is midway between the source position (shotpoint) and the receiving-phone position. This point is called the midpoint. As shown in Figure 24, a reflection point can be the midpoint for a whole family of source-receiver offsets. The traces in that family have one thing in common—the midpoint lies equidistant between their source and receiver positions. Hence, the group of traces has a *common midpoint,* or CMP. If the CMP traces are corrected for NMO and then summed, the resulting *stack trace* has an improved signal-to-noise ratio (compared to that of the individual recorded traces). This happens because each trace in the stack contains the same signal (i.e., the reflection event) that sums coherently, but the random noise doesn't. A collection of traces having a

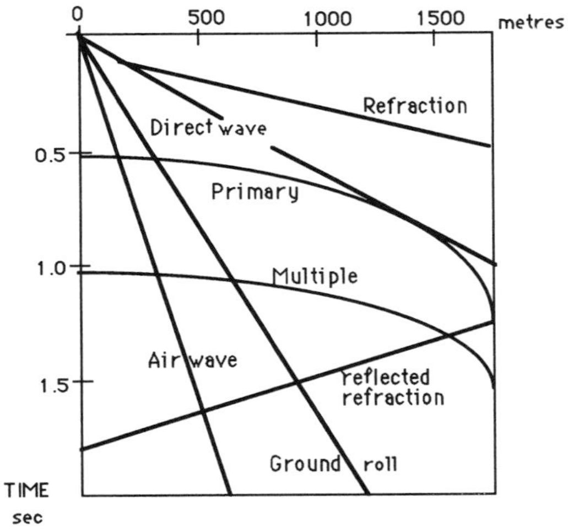

PROBLEM	ATTENUATION TECHNIQUES
Geology induced	
Ground Roll	Source/Receiver arrays and frequency response
	Low-cut filters instruments and phone response
	High number of receivers to trace mix later
	CMP stacking
Multiples	Long offset receivers, high number of stations
	CMP stacking, data processing
Diffraction	Data Processing
Noise	
Wind	Bury geophones, geophone arrays
	High out filters on instruments
	High resonant frequency phones
	Higher energy source
Instruments	Well maintained instruments
	Adequate spares
Power line	Notch filters
	Reverse coil geophones
	Frequency balancing instruments
	Use aluminum foil to cover cables
Aliasing	Spatial-move stations closer together
	Temporal-apply correct high cut filter
Moving traffic	Stop movement during recording
	Sign bit record
	Increase source power
	Vertically stack
Marine Noise	
Cable jerk	Lengthen stretch sections at front and tail-buoy
	Move tow point further astern of vessel
	Check tail-buoy and trailing equipment depth readings
Wave bursts	Shoot through troughs
	Make cable deeper
	Improve cable ballast
	Wait for calmer weather
Cable controller	Move from live sections
	Check for stray flotsam

Fig. 23. Simulated seismic record.

Typical velocities

Refraction 2000 m/s
Reflection 3000 m/s
Sound through water 1500 m/s
Ground roll 500 m/s
Air blast 350 m/s
Direct wave 800 m/s
Multiple 3000 m/s
Shear wave ≈ 1/2 P-wave

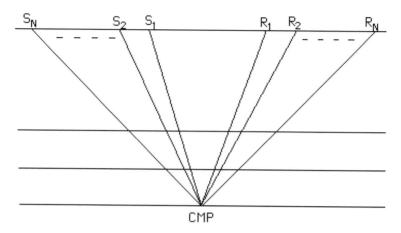

Fig. 24. A common midpoint gather of traces.

CMP is called a *CMP gather*. The number of traces in such a gather is called its *fold*. As for a shot record, the NMO of reflection events in a CMP gather increases as the source-receiver offset increases.

Longer offset data can assist in the separation of multiples from reflection events because the moveout of longer offset reflections is less than the multiple moveout. A longer spread or cable length often improves the process of multiple suppression. This is explained in Figure 25, which shows a schematic example of two primary events and a multiple. At the small offsets, M_2 has a traveltime identical to that of P_2, but at far offsets, the traveltimes differ. The difference in NMO occurs because the two events have different average

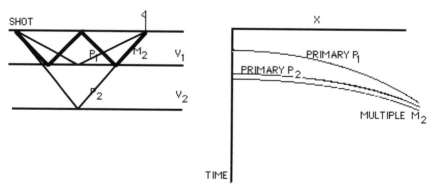

Fig. 25. Multiple reflections on records.

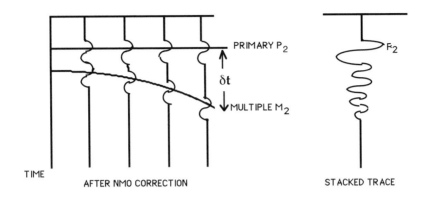

Fig. 26. Stacked trace after NMO correction.

velocities. That is, M_2 travels through only the upper layer with velocity V_1 while P_2 travels through the upper layer with velocity V_1 and the lower layer with velocity V_2. If V_2 is greater than V_1, then P_2 has a greater average velocity. Therefore, according to equation (11), P_2 will have less moveout than M_2.

In data processing, the NMO correction flattens primary P_2 but leaves residual curvature in M_2 (Figure 26). When the traces are stacked, the reflected primary P_2 is enhanced but the multiple M_2 is attenuated. Thus, CMP stacking helps attenuate multiples.

After applying NMO corrections, all traces become pseudo zero-offset recorded traces. If the reflecting beds were flat and the speed of sound in the earth were uniform within beds, stacking the traces would provide an acceptable image of the earth's layering. However, geologic horizons rarely are flat and of uniform sonic velocity. Therefore, the CMP method with simple NMO and stack is not perfect. Nonetheless, data-processing techniques such as *dip moveout* (DMO) are available to account for the departure of the earth from the ideal case (Yilmaz, 1979).

1.4.1 Source/Receiver Configuration and Fold

The evolution of the CMP technique of surveying began in the 1950s. Seismic recording prior to the 1950s utilized a single explosive energy source, firing into a few (six to 12) receivers. One shot would be taken at a shotpoint; then the entire system would be picked up and moved on to another location. Figure 27 shows two often-used survey patterns. One is a *split-spread configuration* with the energy source at the center of spread length X, and the other is

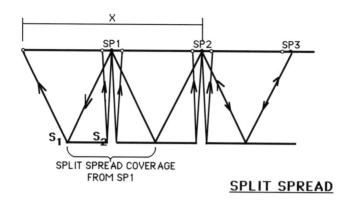

SPLIT SPREAD

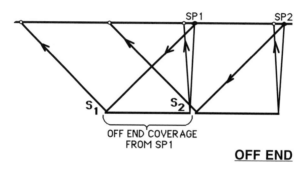

OFF END

Fig. 27. Recording spread configuration.

an *off-end spread* with all of the receivers to one side of the SP. Subsurface coverage from S_1 to S_2 results from a single shotpoint. In noisy areas, ground roll and other noise often masks the reflecting events.

As a result, the signal-to-noise ratio during that era was often extremely low. The technique to improve signal-to-noise ratio, known as the common depth point or common midpoint (CMP) method, was devised and patented in 1956, and publicized in 1962 (Mayne, 1962). With this method, one has duplication of reflections from a single reflection point but with different travel paths. Figure 28 shows how CMP data can be collected by having a shotpoint interval of less than half the receiver spread distance.

Generally, the more traces that are stacked, the more enhanced the signal-to-noise ratio becomes because signal is additive while random noise is not. The fold of a CMP stack is, therefore, an important survey parameter. For particular geologic conditions, an increase in fold above a certain amount may make no noticeable improvement in data quality.

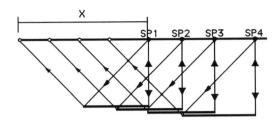

Fig. 28. Duplication of reflection points.

In a CMP survey:

$$\text{FOLD} = \frac{1}{2}\left(\text{No. of spread receiver stations} \times \frac{\text{receiver station interval}}{\text{shotpoint interval}}\right). \quad (18)$$

The value of 1/2 is necessary because adjacent CMP points are separated by one-half the receiver station interval. Equation (18) is valid for a shotpoint interval greater or equal to half the receiver station interval. If the shotpoint interval is less than a half station interval, the number of reflection points increases but the fold of each of them remains the same as that when the shotpoint interval is half the station interval.

Signal processing theory states that the signal-to-noise ratio improvement is proportional to the square root of the fold, if all noise is truly random. However, noise is not always random in practice. Noise can be classified as either coherent (as in the case of a noise train) or incoherent (as in the case of random background noise). Attenuation of noise can be achieved by the following strategies:

Vertical stacking. The shot and receivers remain at the same points, while several separate shots are fired sequentially and recorded. The records are then added to improve signal; random noise that changes in time is attenuated. Coherent noise is not attenuated.

Horizontal stacking (expanded spread). Shot and receivers move away in opposite directions so that they maintain the same midpoints (for later stacking). For land surveys, this strategy has the added benefit of widening the ground-roll-free zone in the records. The method may not be practical, however, in areas having severe topographic changes.

CMP Stacking. Shot and receivers move but individual CMP traces are summed/mixed to attenuate random noise and multiples.

If there is a great deal of random surface noise (such as in high-resolution seismic surveying), the vertical stacking method is probably the best approach. If signal-to-noise ratio is poor because of a weak energy source (or poor reflections), the horizontal or expanded spread method of stacking may be preferred. However, if the energy source is adequate but random noise is still prevalent, the CMP method is preferred.

Equation (18) provides a method of computing the theoretical fold at each CMP location. In practice, the actual fold is often less than equation (18) predicts because excessively noisy or dead traces are removed during data processing. The actual fold at each CMP location (after recording the data) can then be determined by drawing (or having a computer draw) a stacking diagram or chart. This is a fundamental method of checking the fold of coverage of a seismic line.

1.4.2 Stacking Diagrams

Mayne (1962) is accepted as being the first person to propose the use of the CMP recording method to enhance subsurface coverage. To explain its application, he devised the stacking diagram or chart. His charting method will be used here. Such diagrams indicate the amount of fold at each CMP. In the field, stacking diagrams help the geophysicist understand how coverage has been affected by such problems as dead stations and obstacles, which can cause loss of fold.

Stacking diagrams are more important in 2-D land work than in marine work because land source/receiver configurations are variable, whereas the marine source/receiver configuration is always an off-end tow. In 3-D seismic work, stacking diagrams become essential; because of the complexity of 3-D surveys, computer assistance is an essential ingredient in monitoring coverage.

The ends of seismic lines are of particular interest because the CMP technique causes a reduction in fold toward line ends (referred to as *tails*). The tail of a seismic line is referred to in marine work as the *run-out* because the ship must record beyond the end of the desired full-fold line to achieve full fold at that end. In land surveying, the source may continue away from the receiver stations or stop at the end of the receiver line. Either approach modifies the stacking diagram and fold. This is where the stacking diagram helps explain what happens to the coverage value.

There are two methods of drawing a stacking diagram. One method produces CMP lines obliquely whereas the Mayne approach used here produces them horizontally.

Figure 29 shows an example stacking diagram having 12 receiver stations. Here are the step-by-step rules for creating such a diagram:

1) Using a sheet of 1-cm-square or 1-inch-square graph paper, draw each station as a flag along a line at the top of the paper (a flag every centimeter or half inch). Stations must be equidistant.

2) Label flags across the top of the paper; they mark the station positions. In Figure 29, they are labeled as shotpoint 1 (SP 1) onward.

3) Pretend a shot is fired at SP 1 and draw the first raypath down from SP 1 to a convenient reflection point midway between SP 1 and the near-offset receiver position (station 3 in this example). Let the reflection point be on the next square down (1 cm or a half inch down).

4) Draw the raypath from the reflection point back up to the near-offset receiver. Now draw the shotpoint raypaths from SP 1 down to reflection points midway between SP 1 and the remaining receiver stations (4, 5, 6, and so on up to 14 in this example). Draw the horizontal line of reflection coverage through all reflection points. This horizontal line now represents the extent of single fold coverage.

5) Move along to the next shotpoint (SP 2 at station 2) and repeat the exercise to complete the next single-fold coverage line, which is positioned beneath the first SP 1 coverage line completed in step (4). Continue this process for the other shotpoints. The stacking diagram is complete.

6) At any point, count the number of coverage lines that occur vertically under that point. This number is the subsurface fold at that point. For example, station 5 has a fold of 4. The value of fold may be annotated on the stacking diagram where needed.

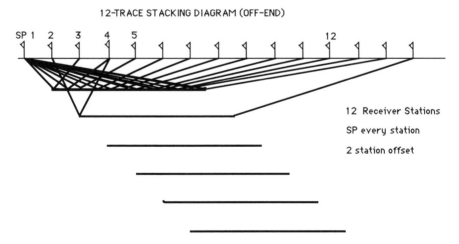

12-TRACE STACKING DIAGRAM (OFF-END)

SP 1 2 3 4 5   12

12 Receiver Stations
SP every station
2 station offset

Fig. 29. A 12-trace stacking diagram (off-end).

The stacking diagram is useful proof of coverage at any point because, if a dead trace exists along the line, a hole will occur in the coverage line for each shot into that dead trace. Dead and unusable traces are not accounted for in equation (18), so the stacking diagram allows the geophysicist to determine quickly which stations have a loss of coverage and the degree of coverage (fold) at any subsurface point.

When the shotpoint interval is every station, the maximum fold is half the number of recorded traces, and hence 6 fold in Figure 29. Maximum fold may be computed from equation (18). If the shotpoint interval is doubled (i.e., every second station in Figure 29), the maximum coverage would be halved (i.e., 3), and so on.

Figure 30 shows a stacking diagram for a split-spread configuration, with the fold value shown at each station location beneath the lines of coverage. Note that with split-spread recording, there is a diagonal gap running down the center of the coverage lines, the width of which is dependent upon the source to near-offset receiver distance. This gap is a loss of coverage caused

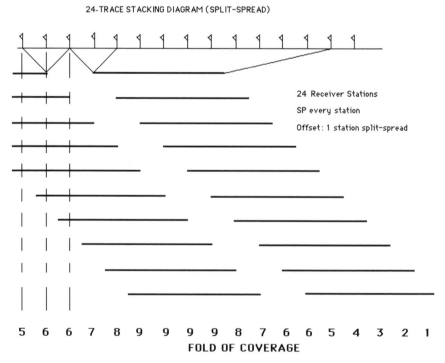

Fig. 30. A 24-trace stacking diagram (split-spread). Not all the traces at the left of the line are shown.

by using a two-station offset in this particular case. If a one-station offset had been chosen, the gap would have been half the width. However, because recording equipment is limited to a set number of recording channels, reducing the near-offset distance (to reduce the gap width) would require moving the whole recording spread closer to the shotpoint. That would result in a loss of the longest offset station. Therefore, survey objectives must be reviewed to determine if trading far offsets for near offsets is reasonable.

Figures 29 and 30 show that at the ends of each seismic line (the tails), the fold reduces from the desired level to zero at the last receiver. To accomplish this operationally in land surveying when using the split-spread recording configuration, the first shot of a line is fired at the first receiver station at one end of a split-spread. With the receivers static, the shots are then fired progressively into the recording geophone receiver spread until the energy source arrives at the center of the split-spread. From then on, the shot and receiver spread both advance (or *roll along*) with each shot. This method allows the fold to build up in value to reach the desired amount at the center of the first receiver spread before commencement of rolling along. (If, on the other hand, shooting had started at the center of the receiver spread, then coverage would begin halfway along the tail leg of the split-spread receiver line). Such a technique, known as *rolling into* or *rolling through* the tail spread, is used to build up the fold at the start of the line. The offset of Figure 30 is one station split, meaning that the distance from source to near receiver on each leg of the split-spread is one station interval.

At the start of a marine seismic line, where the streamer is towed astern, the location of the first shotpoint will become the first full-fold CMP position. The portion of the line prior to that CMP position, where fold gradually increases, is called the *run-in* to the line. If the same technique were adopted at the start of a land seismic line, it would be considered as off-end recording of the tail of the line.

At the end of a land seismic line, the source may be rolled through the stationary spread to obtain the maximum fold at a point half a spread length from the end of line. If maximum fold is required to the end of a survey line, then the geophone stations must be put half a spread length past the end of the line to obtain this fold. For the same reason, a marine vessel will continue shooting for at least half a streamer length past the last station at which full fold is desired.

Today, hand-drawn stacking diagrams, such as Figure 30, are still sometimes used on land crews. The technique shown here is a good learning tool and acts as a good check of any charts drawn on a field, quality-assurance computer. In the marine industry, stacking diagrams are constructed automatically by computer, especially in the 3-D survey case, in which coverage issues are more complicated. In marine recording, hand-drawn stacking diagrams are generally only used in special circumstances (such as in transition zone or erratic coverage areas).

1.5 Survey Design and Planning

If we take a vertical cut through a geologic section, the direction where the geologic units are horizontal is known as the *strike direction*. A geologic section perpendicular to this direction is cut in the *dip direction* (see Figure 31).

The geology of beds is easier to understand if a 2-D profile through them is made in the dip direction rather than in the strike direction. Also, data tend to be of better quality in the dip direction. Hence, dip lines are more important than strike lines in 2-D recording. In 3-D surveying, the situation is somewhat different (see Chapter 7). In 2-D recording, lines shot in any direction other than the dip direction can be confusing to interpret. Consequently, a general idea of basin shape, orientation, or structure initially must be appreciated in order to position lines correctly. In addition, advanced 2-D migration processing is more effective with dip lines and thus a knowledge of the steepest dip direction is of extreme importance in line layout. In a new area to be mapped, seismic lines ideally should be recorded in both the dip and strike directions. The strike lines, in conjunction with the dip lines, help the interpreter form a coherent picture of an area's geology.

Line spacing is determined by the type of survey and the nature of the structure under examination. For reconnaissance work, large line spacing (50 km+) may give a regional picture, and in-fill lines with small spacing (500 m+) may be added later. If an interpreter cannot follow the geologic horizons from one line to the next during his interpretation of the data, the lines are too far apart. In 3-D surveying, the line spacing is required to be as little as 25 m in many cases to provide as detailed a geologic image as possible. Apart from geologic considerations, survey planning cannot proceed until the logis-

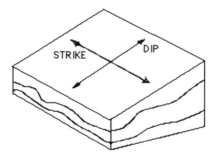

Fig. 31. Dip and strike directions.

tical and economic constraints are understood. A geophysicist should have a broad understanding of the geologic target before a survey is planned and designed.

In planning a seismic survey, the geophysicist must keep the following survey objectives in mind:

Primary: To obtain geophysical data that provide a representation of the subsurface geology that is adequate to meet the interpretational goals.

Secondary: To acquire the maximum amount of data within budgeted funds and time available.

For a particular prospect, a particular acquisition type may well be preferred; for example, explosive sources using single geophone receiver stations may be preferred in marsh conditions. For large structures, wide line spacing using a conventional energy source and low-resolution techniques may be adequate. For smaller structures, closer line and station spacing is required, as well as selection of seismic acquisition parameters that will produce a higher-resolution seismic section.

1.5.1 Seismic Resolution

Structures more complex than the simple dipping structure in Figure 31 are more demanding of the seismic technique. As the search for smaller hydrocarbon traps continues, improved seismic imaging of the subsurface is required. If geology could be viewed in section through a piece of frosted glass, the details observable would depend upon the thickness and clarity of the glass, as well as upon the complexity of the geology. So it is with seismic surveying: The more refined the approach to gathering data, the better the observed geologic image. In the field of optics, if an image is required to be clearer, the imaging equipment needs better *resolving power.* Seismic exploration has borrowed part of this term; the seismic explorer talks of increasing the *seismic resolution.* Thus, more complex structures may require higher-resolution data. This includes both lateral and vertical resolution. High lateral resolution demands close line and shot/receiver spacing (i.e., spatial sampling), whereas higher vertical resolution requires higher temporal sampling of the received data. Good field procedures will produce better signal-to-noise ratio and higher resolution.

1.5.1.1   Seismic Signal Bandwidth

The frequency content of the reflected seismic signal limits vertical resolution. The wider the bandwidth and the higher the frequencies received, the greater the resolution of the final stacked profile and the greater the definition and imaging of geologic beds. If noise is constant in amplitude, then the greater the reflection amplitude and signal-to-noise ratio, the more coherent the final stack. Increased bandwidth also helps lithologic interpretations that depend on detailed knowledge of the amplitude and phase of reflection events. Therefore, it is important that the survey recording parameters do not compromise the survey objectives by either temporally sampling the data too far apart or band-limiting the received frequencies. The seismic signal must be recorded adequately if it is to be reproduced in the computer center at a later date.

1.5.1.2   Target Depth

The target depth is determined by the geologic objective. Acoustic impedance contrasts (i.e., a sonic velocity or density contrast) determine if the target at a particular depth will be a good reflector. The following three issues are closely tied to the depth and nature of the target reflectors:

1) The energy source must have adequate power at the desired frequencies to obtain the target reflections. If the source is too strong, the dynamic range of the recording instruments may be saturated, ruining the fidelity of the data. If the source is too weak, poor signal-to-noise ratio may result and the target will be imaged poorly.

2) There must be sufficient fold to maintain a target signal-to-noise ratio adequate for the geophysicist to make a correct interpretation.

3) The source/receiver geometry must have an adequately long offset to optimize velocity calculations and multiple attenuation in order to resolve the target properly.

Often, survey parameters are optimized for the deepest target, sacrificing the quality and resolution of shallower targets.

1.5.2 Survey Costs and Timing

Economic considerations are important. Without a survey budget, the exploration company cannot sign a contract for a survey to begin. The size of the survey budget often dictates the size of the survey. To make a profit, the service company must work within the contracted price. Budget and survey size are closely related. The geophysicist initially may know the minimum survey size required to adequately image the target, then request the survey costing the least per kilometer from the contractor. An adequate working budget is required to perform the desired survey. A budget to record data in mill pond conditions is useless if the survey is to be recorded in the winter months of the North Sea; a blown budget can be expected before the survey starts. Hence, costs may depend to a great extent on survey timing. The survey budget also constrains the level of survey instrument technology to be used. The lower the budget, the greater the limitation placed on the use of high-cost equipment.

Land acquisition costs are generally higher per kilometer than the cost of acquiring marine data. The cost of both land and marine acquisition is influenced by the mobilization/demobilization costs of the seismic crew to the survey area and the cost of the capital equipment. For large surveys, marine acquisition costs-per-kilometer tend to be lower because, once the vessel is in the survey area, a larger amount of data can be recorded per day.

1.5.2.1 Survey Timing

In selecting when a survey starts, the planner should consider a number of factors:

1) The availability of equipment and personnel. Obviously, a survey cannot begin until the equipment and crew are in place.

2) The recording crew production rate can be a factor in determining when a previous survey may be completed and the next survey starts.

3) If timing is flexible, survey work should be scheduled for good weather periods. That is, with land recording, a dry season is preferred to a wet season (equipment may become immovable and recording may cease during heavy rainfall); in marine recording, it is preferable to record data during calm seas because rough seas may cause safety problems and higher noise, which degrades the signal-to-noise ratio.

4) Permitting is required in all areas of an operation, including permits to gain access to land, permits to store and fire explosives, and permits to use radios. Such permits may take a long time to obtain with land surveys but less time with marine because environmental problems are considered to be fewer in marine operations. Government agencies are often involved in every step of the operation.

5) Visas for expatriate staff are often necessary since some countries apply work limitations on certain visa types. Such restrictions must be allowed for during the planning stage.

6) A great deal of consideration to the terrain over which land recording is to take place may be required in order to determine whether to use four-wheel-drive vehicles, helicopters, or aircraft.

7) Marine operations always require resupply, and it is important to ensure that the ship supplies are ready at the quay side when the ship arrives in port. Otherwise, delays can occur which may compound some of the problems discussed above.

1.5.3 Technical Considerations

Having established the logistical and economic framework in which a survey will be conducted, it is necessary to determine the technical specifications required by the seismic reflection technique. The following technical checklist is intended as a general review of what technical considerations are required prior to the commencement of operations. Later chapters cover these considerations in more detail as indicated in each subsection.

The best way to establish technical specifications is to perform tests in the field. Unfortunately, time and cost to perform tests are often prohibitive; thus, technical specifications of previous seismic surveys in an area often are used. Prior to any survey, *all* previous acquisition parameters and results must be reviewed to see if a change in parameters might improve on previous work. If acquisition parameters are changed, a previous line should be reshot to prove that the newer parameters are an improvement over the previous parameters (assuming there is an adequate financial budget). The key elements of the technical specifications checklist are:

Cable length: Selection is based on a knowledge of moveout of reflections. This is discussed in Chapter 6 in a discussion of parameter optimizing.

Station spacing and minimum wavelength: There must be adequate spatial sampling. An erroneous interpretation may occur when the receiver stations are so far apart that there is ambiguity in the angular orientation of an arriving wavefront. This is termed *aliasing* (just as anyone who has used a false name has used an alias) because a false event may be interpreted. The limiting distance that stations must be spaced to avoid such an occurrence is a function of the reflection velocity, frequency, and dip angle. Thus, station spacing is chosen to avoid spatial aliasing. For further information, refer to a Chapter 6 discussion on channel spacing.

Channel numbers: These are limited to the number of channels the recording equipment can handle. Generally, the greater the number of channels, the shorter the station spacing which can be used in the field.

Geophone Array Design: If a line of single geophones experienced a vertical ground movement together, a lineup of events would appear on the seismic record. This lineup is called a *coherent event*. Because the geophones each provide a single output, any localized noise (such as wind) would generate an output only on the geophones affected. Such signal is considered *incoherent* because it does not produce a lineup of events on the seismic record. If the geophones were connected to each other to form an array of geophones, the array would provide an increased output to any vertical motion as a result of all of the geophones adding their individual output together, whereas incoherent wind noise output from any single geophone would be attenuated.

In the event that harmful coherent surface waves are received (which have vertical motion), the geophones within the array may be positioned so that half of the geophones record the surface wave's peak amplitude displacement while the other half of the geophones record the surface-wave trough displacement. Therefore, with a knowledge of the surface-wave velocity and wavelength, arrays can be designed to suppress much of the coherent surface-wave vertical energy while maintaining output when receiving useful vertically arriving coherent reflections. Geophone array design is discussed in Chapter 2.

Detector frequency: Geophones are tuned to give a peak voltage amplitude output over a desired frequency spectrum of ground motion. Because geophones respond to the velocity of the ground motion, they are known as *velocity phones.* They can be tuned to commence their peak output at an initial ground motion frequency value, from as low as 8 Hz to greater than 100 Hz. This is their "natural" or "resonant" frequency. Geophones display both mechanical and electrical resonance; here, the resonance is the lowest electrical value observed. Because of excessive low-frequency ground-roll amplitude, it may be necessary to choose a higher natural frequency geophone, thereby reducing the possibility of the heavy surface-wave or ground-roll swamping reflected signals.

By contrast, the marine hydrophone is pressure sensitive and gives a flat output across a wide spectrum of frequencies. Further information on receiver design is offered in Chapter 2.

Fold: Generally, greater fold leads to greater signal-to-noise ratio and hence more coherent reflections in the section, although this is not always so. There is, however, a minimum level of multiplicity necessary to produce an acceptable signal-to-noise ratio, and that is considered to be when reflection events are just sufficiently coherent to produce an interpretable stacked section. (This minimum level is commonly utilized in coal exploration where fold is frequently as low as four to six, which produces an adequate seismic image of the coal seams.)

Maximum fold for a given station spacing occurs when the shot interval is a half-station spacing. When the shot interval is shorter than the station spacing, this results in a reduction in the distance between CMP points along a line and does not change the maximum fold value a line can have (see the discussion of stacking diagrams earlier in this chapter).

Line length: This must be longer than the required full-fold coverage length in order that run-in and run-out tails may be recorded to generate the required maximum fold at the end of line points. The geophysical survey plan should allow for full-fold coverage at the ends of lines after data processing (see Chapter 6). A good rule of thumb is to extend the ends of the line by a length equal to the depth of the deepest target.

Energy source: In land surveying, the source with the best signal-to-noise ratio is determined by field experiments. In marine acquisition, there are constraints on the type and size of source that will be used. The geophysicist must be aware of this during the planning stage. Some land sources offer the option of varying energy source output by increasing charge or sweep rate, whereas some marine sources make a similar offer by varying source capacity or gas content (see Chapter 3).

1.5.4 Special Considerations

Because land and marine operations are very different, they each have unique problems and survey aspects that need special consideration by the survey planner. These special topics are described briefly here.

1.5.4.1   Land Surveys

Positioning: This may depend on terrain and production rate. Jungle surveys may require theodolites and flat-plane lasers; in swamp conditions, satellite navigation with beacons and Mini-Ranger equipment becomes necessary.

Statics: Rough terrain, deep LVL, or karst tops may require uphole surveys as well as the use of special refraction crew techniques to obtain data for static corrections.

Noise problems: Traffic (environmental), power line, generator, air, and surface (dynamite blowout) noise may be present to some degree. Specialized equipment may be needed to try to overcome these noise problems.

Crew numbers: A small, high-resolution party typically is comprised of four workers. A normal land crew would be comprised of 40, whereas an Indonesian jungle crew may use 100.

1.5.4.2   Marine Surveys

Positioning: Satellite navigation or offshore range equipment must be used. The choice of equipment depends upon the accuracy required and distance off shore. Land-based ranging systems are used most often, but this may change now that the Global Positioning System (GPS) satellites are fully operational. Round-the-clock GPS coverage still is not available in some parts of the world. Consequently, ranging systems still are commonly used in these areas.

Vessel logistics: Generally, the vessel and crew logistics are described in a survey bid. The length of time a vessel may stay at sea between port calls may be an important consideration. It also may be possible to time share a vessel if the vessel is recording a survey for another company in an adjacent area.

Communications: The vessel is supplied with all the necessary communication equipment installed. It may be necessary to review communications in difficult areas of operation.

Streamer cables: Most seismic ships can tow more than one streamer at a time. For 3-D surveys, costs, data quality, and target illumination are all affected by the number of cables towed. Designers of 3-D surveys must be aware of the tradeoffs involved in selecting a streamer pattern. For 2-D and 3-D lines, the length and group interval of the streamer(s) are important. Also, special consideration must be given to the equipment needed to determine streamer positions during a survey.

Energy source: The energy pulse signature of the source should be known before a survey begins. If possible, perform a *pulse test* prior to survey startup to obtain the source signature for later processing. Alternatives are to use a previous test result or to model the source signature numerically.

Exercise 1.1

During and after drilling a well, a geophysical tool known as a sonde or sonic tool may be run down the well to obtain sonic velocity information about the rock strata. The tool has a transmitter that sends a short sonic pulse through the rock to its receiver. Knowing the transmitter/receiver separation allows a computation of the rock layer or interval velocity. In a similar well-logging run with a density detection tool, relative rock densities are determined. Knowledge of interval velocity and density allow a crude form of seismic modeling.

Figure 32 represents a well drilled through four dipping layers. R is the reflection coefficient at each boundary, and ρ and V represent sonic density and velocity, respectively. Four interval velocities have been determined for each of the four beds; a density log is also available. Values are provided in Figure 33. Draw the seismic model representing the well, in the form of a "stickogram," with the reflection coefficient R_c being represented by a stick and having maximum deflection amplitude values of ±1. Ignore traveltimes. The stickogram would have positive *or* negative reflections at levels R_1, R_2, and R_3, as indicated below.

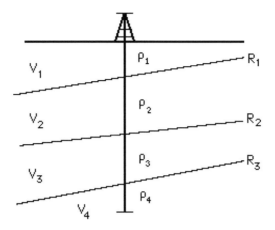

Fig. 32. A well drilled through four dipping layers.

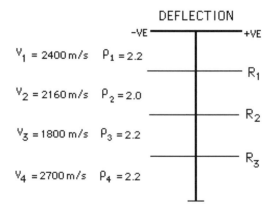

Fig. 33. Values of four interval velocities.

Exercise 1.2

Draw the stacking diagram for a 12-channel split-spread recording config-uration, in which the station interval is 10 m, the shot interval is 10 m, and the near-offset distance is one station. Draw the diagram for 10 shots in which station 6 is dead throughout the recording, and label the fold at the base of the diagram.

Chapter 2

Receiver Design and Characteristics

In geophysical exploration, seismic data are acquired by firing an energy source on or near the Earth's surface and recording the energy reflected back to the surface from the geologic substrata. This chapter discusses the methods by which the geophysicist detects reflected energy on land and offshore. Geophones are used by land explorers, and hydrophones are used in the marine exploration industry. The energy detected at the surface contains useful signal but also unwanted noise.

This chapter explains the operation of the receiver phone, how it converts acoustic energy into an electrical signal, and how this signal is passed along cables to recording instruments. Because much of the useful reflected energy is weak compared with geologically generated coherent surface noise (see Chapter 1), the geophysicist often must attempt to enhance the signal level and reduce the coherent noise level. We exploit the fact that the useful signal is arriving almost vertically when compared with the horizontally arriving noise. The signal is enhanced, when compared to the noise, by using many receivers in a geometric pattern that attenuates energy traveling horizontally. Such arrangements are known as *arrays*.

A complete discussion of array theory requires the reader to have a good knowledge of mathematics; since that is not the intention of this book, the mathematical treatise has been included as Appendix B. In contrast, this chapter describes a simple, practical approach to array design and noise attenuation that the reader can apply to real field problems with the aid of a simple calculator. However, the reader should keep in mind that array theory assumes that all geophones have identical coupling with the earth, that the phones are all positioned at the same elevation, and that there are no errors in horizontal positioning. This is never so in practice, which means that actual field arrays never reach the theoretically expected performance.

47

2.1   Land Receiver Systems (Geophones and Cables)

Conventional geophones are based on *Faraday's Law of electromagnetic induction*. This law states that relative motion of a conductor through a magnetic field induces an electromotive force (EMF) which causes a current to flow through the conductor if the conductor is an element of an electrical circuit.

Figure 34 shows a cutaway view of a simple geophone and a simplified schematic sketch of the main parts of a geophone. The schematic is useful for understanding how the geophone works. The case that houses the geophone element is shown as being in rigid contact with the earth and a magnet. The essential ingredients to make a geophone are a permanent magnet, a conductor, and a spring, which positions the conductor in the magnetic field space. Schematically, the magnet is shown with north and south poles; the lines of force pass between the two poles. The conductor is represented as a wire passing between the poles; it is shown to be connected to an external electrical load by the resistor at each end of the conductor.

The conductor, in reality, is a length of copper wire wrapped into a cylindrical coil shape. It is often referred to as the *coil* or *element*. This coil has mass, which dictates how much strength the springs must have to hold the coil between the magnet's poles. Hence, the schematic shows this mass, suspended in space, as an addition to the electrical coil. Only the coil is affected by the magnetic field, so only it is placed within the field.

If the earth moves upward, the magnet and case must follow because they are in rigid contact with the earth. Initially the mass will try to remain in the same place because of its inertia or move down with respect to the upward-moving magnet and frame. Then the spring will exert an upward force on the mass, making it move upward. When the earth and case stop moving upward, the inertia of the mass causes it to continue upward until pulled to a stop by the lower spring. Hence, the mass motion lags the case movement. If the case is not moved back downward, the lower spring will pull the mass down until the upper spring stretches so far that it stops the downward motion; then the mass starts moving upward again. Thus, the initial motion of the case causes the mass to oscillate between the two springs. Since energy is constantly being dissipated through the springs and by the mass movement through air, the mass movement gradually decreases in amplitude. Physicists call this behavior decaying *simple harmonic motion* (SHM).

The conductor's motion through the magnetic field, according to Faraday's Law, causes an EMF to be induced that is proportional to the velocity of the earth's motion. Hence, such a geophone is called a *velocity phone* because its output is proportional to the velocity of the earth's motion. Since the coil

(a)

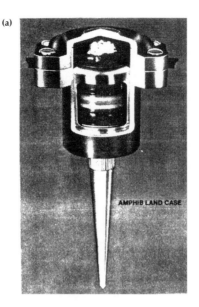

(b)

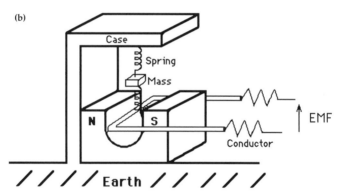

Fig. 34. Geophones. (a) A geophone cutaway (courtesy of Mark Products). (b) A schematic diagram of a geophone.

and magnet are always engineered as a single package within the geophone case, this package is often referred to as the element.

2.1.1 Frequency Response and Damping

If the mass is displaced from its equilibrium position then set free, it will experience SHM. This is called "undriven" SHM because no outside forces are driving the motion. The output frequency and voltage of an undriven geophone are given by equations (19) and (20):

$$f_n = \frac{1}{2\pi}\sqrt{\frac{K}{M}}, \tag{19}$$

where f_n is the natural or *resonant* frequency, K is the spring constant, and M is the mass of the geophone; and

$$E = BLVc, \tag{20}$$

where E is the output voltage EMF, B is the magnetic field strength, L is the number of turns in the coil, V is the relative velocity of the coil with respect to the magnet, and c is a component geometry factor set by the manufacturer after laboratory testing of the geophone response (using coils of different area and material).

When seismic reflections arrive at the Earth's surface, the earth moves up and down at different velocities, causing the case of the geophone to move up and down. The geophone is then said to be "driven," and the output is no longer SHM. As the case moves, so the mass moves later in time, dependent on the strength of the spring constant. This time lag between the case motion and the mass motion is termed the "phase response" of the geophone. The relationship between the coil velocity V and the ground velocity is given by equation (21):

$$V = \frac{D\cos\phi}{R}, \tag{21}$$

where D is the vertical ground velocity, ϕ is the phase between the case motion and the mass motion, and R is the damping constant.

Because of its electrical and mechanical characteristics, a geophone responds differently to different excitation frequencies. Ideally, a geophone should give an equal voltage amplitude output for equal coil displacement across all frequencies. This is called a *flat response* across the frequency spectrum. Since in practice the output of a geophone is not the same at all displacement frequencies, the geophone will distort the true shape of the sound waves it detects. Also, an ideal phase response should be zero across all frequencies. This is extremely difficult to achieve in practice because of variations in spring materials and mechanical coupling variations of the case with the earth at different frequencies.

Figure 35 shows the amplitude response of a typical geophone as a function of output voltage damping. Clearly, using the natural response (23% damping) will incorrectly emphasize frequency content at the geophone's natural frequency (14 Hz in this example). If a resistor is placed across the geophone coil's output, some energy will dissipate through the resistor and reduce the output. This is known as *damping* the coil's output. The damping resistor is also called the *shunt resistor*. The lower the resistance value used, the greater the amount of damping.

An undamped geophone has a limited amount of damping (23% in this case) because of the electrical resistance of the coil. The damping of the geophone is also affected by connecting that geophone to others and by the impedance of the recording system. In today's systems, the overall impedance is considered negligible.

If a geophone has no shunt resistor (shunt open implies an infinitely high resistance) and is tapped lightly, it will oscillate for some time. As the shunt resistor is decreased in value, the number of oscillations will decrease because damping increases until a point is finally reached where a tap will fail to produce an oscillation. This is the point at which the geophone is critically

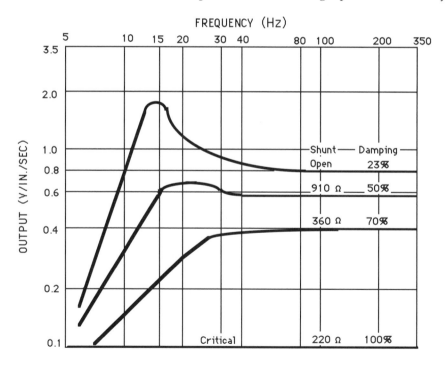

Fig. 35. Geophone response.

damped. The various curves shown in Figure 35 correspond to different values of damping and are typical of a 14 Hz geophone.

The different values of damping are expressed as a percentage of critical damping. As the damping increases, the output peak in the frequency response decreases in amplitude and moves toward higher frequencies. The peak disappears between 50% and 70% of critical damping, at which point the flattest frequency response is obtained. The value of the shunt resistor used is determined by laboratory tests to ensure that a geophone's output is the desired response characteristic required by the manufacturer.

Generally, geophones are manufactured to have a natural frequency in the range of 1–60 Hz, but most commonly 4–15 Hz geophones are used for oil exploration reflection work. For shallow reflection work, 100 Hz and 400 Hz geophones are used sometimes, whereas 1–10 Hz geophones are used in refraction work.

2.1.2 Electrical Characteristics

2.1.2.1 Sensitivity

Modern geophone construction ensures high sensitivity compared with the maximum sensitivity of geophones of yesteryear. Geophones are available with a wide range of sensitivities. For example, at one end of the sensitivity scale, a Mark Products geophone can produce 0.1 V output for a 2.5 cm/s (1 inch/s) velocity, while another geophone can provide as much as 0.4 mV output for a tiny movement of 2.5×10^{-8} m/s (1 μ inch/s).

2.1.2.2 Tolerances

Geophones have typical tolerances as follows:

- Natural frequency within ±0.5 Hz of the manufacturer's stated value
- Natural frequency distortion with a maximum 20° tilt, ±0.1 Hz
- Sensitivity within ±5% of the manufacturer's stated value

To ensure tolerances are maintained, geophones should be checked independently at least once a month. This can be done on a shaker table or with an electrical impulse from the recording instruments or separate instrumentation. The simplest method to determine if a geophone is functioning at all is to use a voltmeter and manually shake or tap the geophone.

2.1.3 Physical Characteristics

Figure 36 shows different types of typical geophones and their cases. A geophone couples with the ground through a spike screwed into its case.

Geophones intended for different applications have different case types. The land case has wiring connections on each side for ease of handling and stability. The marsh case has top wiring connections and a slim vertical case for deeper planting, with a broad spike for soft spongy material. The spike must be long enough for good coupling in hard ground but short enough to allow the case body to touch the ground. Generally, it has been found that the longer spike gives improved amplitude response. Flat bases are sometimes used over rocky terrain where it is too hard to put spikes in the ground, but such poor ground coupling usually produces inferior data quality.

A gimballed version of the geophone allows the element always to be oriented vertically. These geophones are used in bay and shallow marine areas where obstacles on the bottom prevent the case from sitting upright. They also can be used in operations in snow-covered terrain.

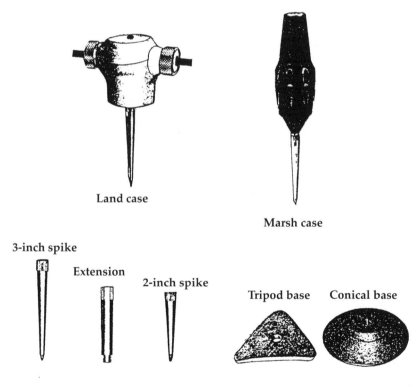

Fig. 36. Geophone types.

2.1.3.1   Reliability

Geophones do not give their specified output when they lie on their side, are open circuit, or have faulty circuitry. Faulty damping resistors can cause sensitivity and phase distortion. Failure rates may be as high as 10% per month, particularly with new geophones. Rugged use in harsh environments increases wear and tear on mechanical moving parts that can affect a geophone's output frequency and phase response.

2.1.3.2   Harmonic Distortion

Geophones have nonlinear characteristics because of imperfections in the mechanical and electrical components. For example, the relative motion of the coil through the magnetic field is not a perfectly straight line because of irregular spring forces. Also, electrical distortion prevails during extreme or rapid coil displacements. Geophones generally are manufactured to reduce such distortion. When quoting distortion levels, the supplier should indicate the geophone driving frequency and velocity for the measurement and state that the value provided is for total harmonic distortion, not just a single harmonic. For example, one commercially available geophone has less than 0.1% total harmonic distortion at resonant frequency with a velocity of 1.78 cm/s (0.7 inch/s) per second.

Manufacturers's literature sometimes refers to "digital grade" geophones. Such a geophone is made for use with digital recording equipment for which distortion can be tested, whereas such tests could not be performed with the older analog equipment that used lesser-tolerance geophones. Because the industry uses only digital recording equipment today, the phrase is no longer relevant.

2.1.3.3   Noise

In addition to external environmental noise sources (wind, traffic, vibration), a geophone is sensitive to two kinds of internal noise: thermal noise as a result of coil resistance, and Brownian particle movement within the case body. Because of the nature of thermal and Brownian noise phenomena, there is no easy solution for removing them. However, the internal signal-to-noise ratio caused by the coil's movement within the geophone element is proportional to the square root of the coil's mass. The element is therefore designed to minimize such noise.

Other background noise, known as *electrical pickup*, results from external influences such as electromagnetic fields (power lines), Earth ground systems, cathodic protection of pipelines, thunderstorms, charged fences, shot blasters, and radios.

To attempt to reduce pickup, the geophone impedance is kept low by reducing shunt and coil resistance values and by modifying the magnet design accordingly. The plastic geophone body insulates the geophone to a degree. Dual coils have been tried (with limited success) to give isolation from electromagnetic fields such as power lines. With dual-coil geophones, two coils are physically arranged with opposing magnetic fields so that their overall field is zero, while their respective output is summed. Dual coil geophones are not used routinely because the industry found a better solution—phase-canceling *balancing boxes*—to the power-line noise problem. Such boxes cancel the recorded power-line frequency by summing an input voltage of opposite polarity to the power-line frequency at the input to the instruments. However, when power-line frequency is not stable, such balancing boxes become useless. (The author has observed frequencies as far off as 5–7 Hz from the expected power supply frequency, particularly where local farm generator sets are used.)

2.1.4 Special Geophones

A geophone will give an output only when it has undergone motion along the axis of its element. That is, the geophone will give an output when moved vertically; this output diminishes rapidly as the phone is tilted. A phone will not give an output when motion is horizontal to it. A geophone on its side will not respond to vertical waves, if it responds at all. Thus, shear (*S*-wave) geophones are built for the purpose of recording horizontal particle motion. A level bubble aids horizontal placement, so planting of such geophones tends to be slower since more care is exercised than with vertical geophones (Figure 37).

Manufacturers are upgrading their geophone designs continually. One

Fig. 37. Shear-wave geophone (courtesy Mark Products).

company offers a shear-wave geophone that is gimballed in a lubricating fluid so it will perform well at tilt angles up to 30° (Figure 38). The company also offers a three-component geophone that uses a bubble level indication of ground positioning (Figure 39). Three-component geophones having elements mounted at right angles may be used where both S- and P-waves are to be recorded. In the field, such a geophone must have the bubble centered to

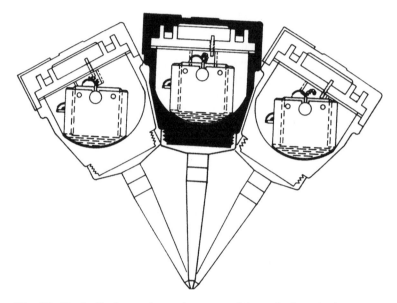

Fig. 38. Gimballed geophone (courtesy Litton Industries).

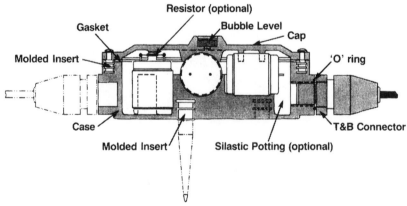

Fig. 39. Three-component geophone (courtesy Litton Industries).

ensure all three components are correctly aligned horizontally and vertically. Two companies offer a geophone in which the moving part is the magnet; the fixed part is the coil. One company claims its geophone allows 100% operation irrespective of tilt, and hence the same geophone may be used for either *S*- or *P*-wave study.

2.1.4.1   Transition-Zone Phones

Transition zones are areas where the sea meets the land such as shallow water coasts, beaches, shallow reefs, or river estuaries. Data acquisition is very difficult in these areas. One approach adopted for recording data is the use of thick-skinned cables (marine cables built using multiple cable skins), where thin-skinned cables would otherwise rip on rugged seabeds and reefs. For many years, these have been used (and still are in use) in shallow water *back-down operations*. The cable is known as a *bottom-tow cable*, the operation being simply to tow the cable toward the desired shotpoint, and when correctly positioned to back the towing vessel down (thereby allowing the cable to sink to the bottom) and then record the shot.

These thick-skinned cables are also used for inshore shallow-reef operations, which is an acceptable practice where there is a fairly constant layer of water covering the cables, allowing the hydrophones to produce an output. However, such cables are often ripped by incessant wave action, causing the recording of poor data of low signal-to-noise ratio. Instead, we may use special geophones (similar to marsh phones) that are made with long spikes for insertion into sand. Such phones are built to be moisture resistant so that they may be used in beach or tidal flat areas where they are covered with a shallow water column for a limited period of time (see Figure 40). Such phones are fre-

Fig. 40. Transition-zone recording.

quently submerged, but their cables may be difficult to handle physically under surf conditions. To overcome this problem, phones have been made which have a radio transmitter that transmits the output back to the recording instruments (known as *telemetry*) from a whip antenna mounted on a buoy built to float (Figure 41). The geophone may then become the buoy's anchor to the sand or seabed. Some form of alternative anchor is sometimes used to remove the strain from the geophone cables to prevent wear, tear, and vibration on the phone cables. On tidal flats, when the tide regresses, the unit housing the antenna falls on its side, and radio transmission may be affected. Therefore, this type of operation has its limitations.

Wave noise presents a serious problem when surf crashes onto each phone; on beaches, geophones may be buried in the sand in an attempt to reduce such wave noise, but the noise from a pounding wave is almost impossible to attenuate. Since waves have a variable wavelength and possess a largely vertical motion as they roll along the shoreline, noise cancellation by the use of arrays (which will be discussed later in this chapter) is of little use. Even if the phones in Figure 40 were arranged in some form of array, the waves would pound against the connecting cables, thereby vibrating the geophones to produce an output. Consequently, transition-zone data are often of better quality during days when the sea is calm.

A geophone for use in areas where the water depth is less than 200 m may have a hydrophone built in alongside the geophone. These are known as *dual-sensor phones* (Barr et al., 1990) and may be built into a bottom-tow cable. The

Fig. 41. Shallow water buoys (courtesy Digiseis).

geophone is gimballed in one direction to allow it to maintain an upright position because the cable may settle on an uneven seabed. The relative impedances and mechanical responses of the two phone types are matched so that frequency and phase differences are minimized. The benefit of such dual-sensor cables is that the two types of data can be combined so as to eliminate the water column multiple or *ghost*, discussed in Section 2.5.

A geophone and a hydrophone are placed together on the sea floor and connected to the recording system so that their responses to an upward-traveling sound wave are in phase. Because a geophone's response is dependent on direction, but a hydrophone's isn't, downward-traveling sound waves produce two responses that differ by 180-degree phase. If these two phone outputs are properly scaled and then summed, all sea-air reflections are canceled and, hence, water-column multiples are attenuated. This summation does not affect the upgoing arriving reflections since both types of phones respond to them with the same polarity output. Current research is focusing on using multicomponent geophones in ocean-bottom, dual-sensor acquisition.

2.1.5 Geophone Response Testing

The response of a geophone can be determined either mechanically or electrically (by an impulse method). Figure 42 indicates the mechanical method, which uses a shaker table. A calibrated coil is used as the reference geophone and is coupled mechanically to the test geophone. A resonator vibrates both geophones vertically, and their electrical output may be displayed on an oscilloscope as a Lissajous figure, as shown in Figure 43. If the geophones are identical, a straight line at 45° will result. If the phones have

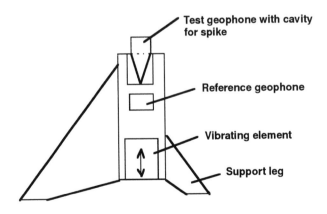

Fig. 42. The shaker table mechanical schematic.

different amplitude responses, the slope of the line will change. If they respond at different frequencies, the line will expand into an ellipse.

The impulse method (Figure 44) is performed by providing an electrical spike into the geophone and by determining the resultant coil electrical amplitudes and crossover delay times as the geophone responds. Thereafter, the parameters of the geophone may be calculated. Clearly, the impulse method can test any geophone type, whereas the vertically vibrating shaker table is only useful for vertical geophones. (Horizontal shaker tables are used rarely.) Although the impulse equipment is less bulky than a shaker table, the method does require some analysis of the geophone output signal.

When a geophone is shaken, its coil motion produces an electrical signal across its output terminals, as explained earlier in this chapter. If a geophone has an electrical signal applied across those same terminals, the signal is said to "drive" the coil, causing the coil to move in its case. When the driving sig-

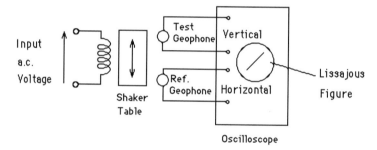

Fig. 43. The shaker table electrical schematic.

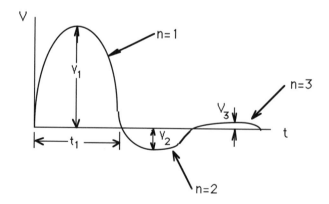

Fig. 44. Geophone response to a step acceleration.

nal is in the form of a positive electrical spike, it causes the coil initially to move upward. With no further signal applied, the coil will drop and continue moving up and down, depending on its damping and electrical output characteristics. If the phone's output is wired across an oscilloscope, the output voltage would appear as shown in Figure 44. Using this information, the geophone's resonant frequency may be computed from equation (22). Such a test is normally performed in a laboratory (Aston, 1977).

Consider

$$f_n = \frac{1}{2t_1\sqrt{(1-d^2)}}, \tag{22}$$

where f_n is the natural or resonant frequency, t_1 is the time of the first zero crossing, and d is the damping constant. The damping constant is a function of the geophone voltage ratio, where $r = V_n/V_1$ (see Figure 44).

Typically, $n = 2$, but sometimes V_2 is difficult to measure. In that case $r = V_3/V_1$, and $n = 3$ may be a better choice for the voltage ratio.

Use

$$d = \frac{-\log_e r}{\sqrt{(n-1)\pi^2 + \log_e^2 r}}, \tag{23}$$

for coils with damping between 0.2 and 0.8 (dimensionless units) of critical damping. For coils with less than 0.2 of critical damping, V_2 can always be measured, so

$$d = \frac{-\log_e r}{\sqrt{\pi^2 + \log_e^2 r}}. \tag{24}$$

For a perfectly undamped system,

$$d = 0, \text{ and } f_n = \frac{1}{2t_1}. \tag{25}$$

2.1.6 Cables

The connection of the geophone string to the recording cable is known as the *string takeout* and the connector is the *clip*, whereas the recording cable connection to the data-acquisition system is the *cable takeout* and the plug on the end of each cable is known as the *cable head*. These electrical takeouts are designed so that their connection clips mate in only one direction, thus preventing an accidental polarity-reversing connection. Geophones are connected to each other and to the main multiwire recording cable via a twin pair of wires (one pair is used as output to the cable while the other pair may be used for connecting phones in a serial/parallel configuration), which is referred to as a *geophone string*. During transportation and storage, geophone strings are attached with wire or nylon loops to a hasp or *pin*. Methods of handling multiple geophone strings are shown in Figure 45. The metal hasp is carried by hand.

Types of cable takeout connectors are illustrated in Figure 46. Takeout clips can be paralleled, as shown in Figure 47, to provide more geophones per recording station.

Takeouts suffer from all the normal problems associated with electrical conductivity; that is, they can be shorted (by wires touching each other or moisture causing shorting between wire pairs) or open circuited (by wires

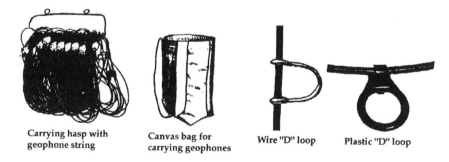

Carrying hasp with   Canvas bag for
geophone string   carrying geophones   Wire "D" loop   Plastic "D" loop

Fig. 45. Carrying equipment.

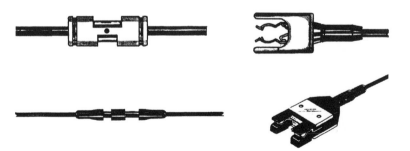

Fig. 46. Cable connectors.

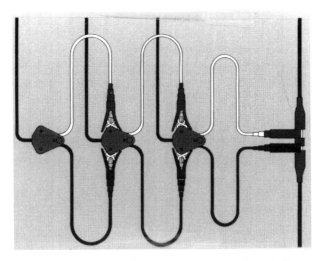

Fig. 47. Parallel clips (courtesy Litton Industries).

breaking as a result of rough handling). They also can experience electrical crossfeed (as a result of moisture or poor wire insulation).

An ideal recording cable should have a small diameter, a low resistance, be of low cost, be reliable, be capable of handling a large number of channels, be lightweight, be simple to repair in the field, and not suffer from normal electrical conductivity problems. Unfortunately, these features are not always compatible and tradeoffs have to be accepted. Copper wire is often used rather than an alloy because thin wire conductors reduce the cable bulk and many alloys cannot take the same strain as copper. Cables are constructed in various lengths dependent upon the manufacturer's requirements of distance between and number of geophone takeouts.

Fiber-optic cables using multiplexing of signals achieve many of the ideal objectives mentioned above, but much fieldwork continues to be done with bulkier multipair-wire cables. Fiber optics provide minimal resistance, leakage, pickup, and line losses. The problem with fiber-optic cables is that they don't easily lend themselves to field repair, as do normal wires that can be soldered. Also, since the electrical signal must be converted to light in the field, this means that light-converter equipment (powered by batteries) also must be distributed in the field with the geophones, which can add extra work and expense to the survey.

2.2   Marine Receiver Systems (Hydrophones and Streamers)

Where geophones and cables are used as the receiving equipment on land recording, so hydrophones and their cables (streamers) are used in water. Solid structure supports elastic wave propagation, but liquids support only acoustic propagation. This means that, in land recording, both compressional and shear waves are received as a physical movement of the geophone. In the marine case, however, only a compressional wave can be transmitted through the water. The pressure changes caused by compressional waves are detected by hydrophones.

2.2.1 Hydrophones

The hydrophone is an electro-acoustic transducer that converts a pressure pulse into an electrical signal by means of the piezoelectric effect. If mechanical stress is applied on two opposite faces of a piezoelectric crystal, then electrical charges appear on some other pair of faces. If such a crystal is placed in an environment experiencing changes in pressure, it will produce a voltage proportional to that pressure.

The piezoelectric crystals are sealed to prevent water ingress and hence electrical leakage. Various styles of hydrophones are manufactured, but generally the hydrophone consists of two crystals, each shaped in the form of a – 2-mm-thick, 2-cm-diameter disc, housed in a thin alloy container. The housing is about 4 cm long, 2 cm in diameter and is designed to allow the transfer of the pressure field through the metal housing and apertures in its body. This housing is also the mechanical junction point for wires going to each crystal disc.

The responses of a hydrophone crystal to acceleration due to unsteady streamer towing and pressure due to seismic signals are explained with the aid of Figure 48. For simplicity of explanation, the crystals are shown with front and side views and the direction of tow is right to left. During streamer tow, the crystals' two sides are parallel with the direction of tow and the physical compression is therefore along their axis as shown in the upper front-view figure. This is schematically represented in the upper-side-view figure. Compressional waves arrive at approximately right angles to the direction of tow (or some component of compression will) as indicated in the lower-front-view figure. Such an arrival has the effect of physically compressing the crystals as shown by the dashed lines in the lower-side-view figure.

By summing their positive and negative electrical output as shown in Figure 49, the output as a result of acceleration is canceled, whereas it is summed during compression. Thus, the crystal pair performs as a single acceleration-

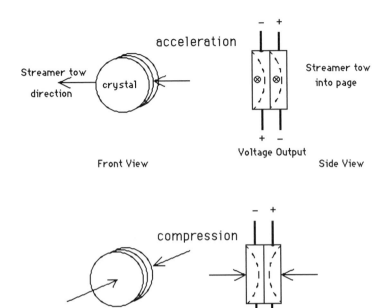

Fig. 48. Hydrophone response to acceleration and compressional waves.

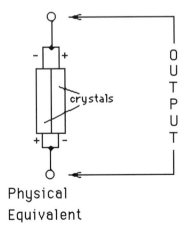

Fig. 49. Hydrophone connections.

canceling hydrophone. The hydrophone is referred to as acceleration canceling because it reduces tow acceleration noise while providing an output for
compressional arrivals. This helps to provide a signal-to-noise ratio which
meets industry standards for streamer operation.

Most hydrophones fall into one of three categories: flat disc, separate disc,
and ceramic. A single hydrophone's sensitivity is typically in the range 25-
48 μV per μbar (a bar being equal to 14.5 psi, which is approximately 1 atmosphere at sea level). Hydrophone pairs are summed in series to improve their
voltage output response. This voltage is modified by a transformer, which
increases the current into the recording instruments. The typical series-connected hydrophones shown schematically in Figure 50 give a transformed
sensitivity of 5 μV per μbar, which is an acceptable value for the recording
instruments. Such a set of hydrophones is built as an array of hydrophones. It
is also conventional to refer to the array as a *group* with any number of groups
up to about 240 making up the cable, or streamer. Streamers using up to 480
groups have been used, but the more common figure in use is 240.

The pressure amplitude of a seismic wave is linearly proportional to its
frequency. Since a hydrophone's output is a function of that pressure, the output voltage varies with the frequency of the seismic wave. This is undesirable
because output will therefore increase at higher frequencies and, just like the
geophone, we want a flat frequency response. A damping resistor is therefore
placed across the array to flatten the response as shown in Figure 50.

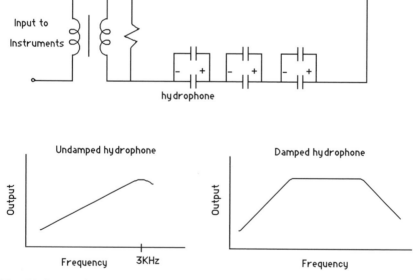

Fig. 50. Equivalent circuit for a streamer group.

A customary quiet sea state generates a background noise of about 1 μbar. Thus, a typical hydrophone array will produce about 5 μV of noise at the recording instrument's input. In normal field operation, the sensitivity is accepted as the manufacturer's quoted value, and only when a shot is fired can comparison be made with other groups. However, if a calibrated hydrophone is available, sensitivity may be checked (and may be found to be 10% less than the manufacturer's specification after three months of use because of natural wear of the crystals caused by constant buffeting by the sea while under tow and frequent handling as the cable is reeled in and out).

2.2.2 Streamers

The first streamers (so called because they "stream" behind the towing vessel) were land cables with hydrophones taped to the cable's side. Buoys maintained the cable at the required depth at chosen points. High noise was generated because of vortex action under tow. The catenary between suspension points resulted in nonuniform streamer depth (Figure 51).

Later (the 1950s and 1960s), a streamer cable was designed for naval (submarine and mine detection) operations; this basic design is still in use today (Figure 52). A streamer cable is formed by connecting together subunits called *sections*. Each section contains one or more hydrophone groups.

A section is typically 12.5 to 100 m long with end connector couplings. Each section may have 15 to 100 hydrophones connected in groups to form two to eight receiver stations. Steel-wire stress members join plastic bulkheads together, and the hydrophones are suspended between the bulkheads in a light kerosene oil called *naroma*. The bulkhead protects the hydrophones from excessive pressure and wear whenever the streamer is reeled aboard the vessel, provides a mechanical coupling between the steel stress members, helps isolate one part of the streamer from the next, and tends to reduce noise traveling through the streamer. Each streamer section is housed in a clear plastic (PVC) sheath or *skin*. A typical 3000 m length streamer can weigh a few tons; after filling with kerosene, the streamer can weigh as much as six tons. When traveling from one prospect to another, the streamer is stored on a large drum or cable *reel*.

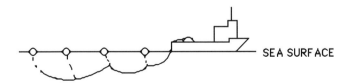

Fig. 51. Catenary between suspension points in early streamers.

Because kerosene is lighter than sea water, a streamer filled with enough kerosene to counter its dead weight can be placed at a particular depth and will remain there. This is known as *neutral buoyancy*. When filled with additional kerosene, each section will become positively buoyant, which can be an advantage if the cable is accidentally cut (causing it to float to the surface for retrieval). If sea salinity conditions change or an excessive amount of kerosene is put into the streamer, strips of lead weight may be taped to the PVC skin to make the section neutrally buoyant again. However, such external weights are undesirable because one positioned near a hydrophone may generate excessive vibration noise. Ideally, the cable should be neutrally buoyant such that other mechanisms may guide the streamer to the required depth without buoyancy problems.

A typical streamer system is illustrated in Figure 53. The lead-in section has a solid core rather than being filled with kerosene. A flexible metal sheath provides protection against damage from vibration and rubbing where the lead-in contacts the ship's stern pylons, back deck, and cable reel. The lead-in also must be capable of withstanding high pressure from the energy source because it may come in contact with the energy source equipment during vessel turns.

Not every section in a streamer has to contain hydrophones. Sometimes *dead sections* are used to achieve desired group spacings. Such dead sections contain no hydrophones and are interspersed between the active or *live* sections. Combinations of dead and live sections provide a great deal of flexibility in designing optimum receiver arrays.

The principal purpose of the tail buoy is to provide a reference on the cable position. This is important at the beginning of a line to ensure that the cable

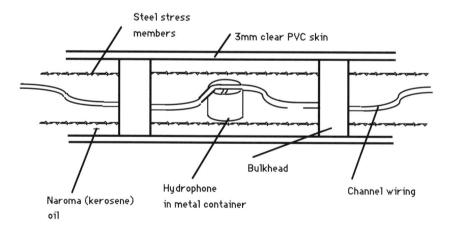

Fig. 52. Marine streamer schematic.

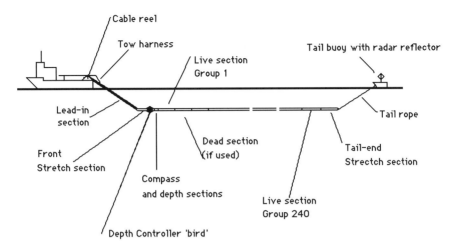

Fig. 53. Marine towed configuration.

has straightened out after any pre-line ship maneuvers. During shooting, it may also help indicate cable drift as a result of crosscurrents. The tail buoy has a radar reflector and may have a strobe light mounted on it. A bearing on the buoy can be obtained on the ship's radar or, with less precision, by observing the buoy through binoculars. Rough seas can make it difficult to locate the buoy because of its limited size and height. In some cases, a low-power radio beacon is mounted on the buoy. Automatic direction finding (ADF) equipment on the vessel may be used to give a continuous bearing to the radio beacon. The tail buoy serves as a marker for ships passing astern and also aids in retrieving the cable if it is cut.

One or more stretch sections are installed behind the lead-in to attenuate cable noise caused by the vibration and jerk from the ship. The stretch section has a vinyl jacket and often contains kerosene. Rope is used in the stretch sections in place of the wire stress members to give a stretch capacity of up to 50% of the relaxed length. Any section that is stretched more than 20% should be tested before further use.

2.2.3 Depth Control

The depth of the streamer is controlled by attaching winged devices called *birds* to the cable. On passive birds, the wings (sometimes called "diving planes") are controlled by spring tension preadjusted to make the bird *fly* at the target depth, typically about 11 m. When the cable reaches the target depth, the water force on the wings will be balanced by the spring tension. If the cable should drop to a lower depth, the angle of the wings will be reduced, causing the cable to return to the target depth. Conversely, a decrease in cable depth produces a greater angle of the wings, causing the bird to seek a deeper level. This type of system also is affected by vessel speed. By varying the speed of the vessel, say between 4.8 knots and 5.2 knots, some overall adjustment of depth is possible without retrieving the cable to reset the spring tension. However, this also can be a disadvantage if it dictates a vessel speed that is unacceptable for reasons such as noise, cable stability, or desired production speed.

A better system of control is with birds containing active control elements (Figure 54). The wings are attached to a water-piston, tension-spring mechanism that causes the assembly to seek a depth to which the spring tension has been programmed. The desired depth is controlled individually by a command transmitter (also shown in Figure 54) in the ship's instrument room. The observer sets the depth on the controlling computer, which then transmits a pulse along the streamer. The bird recognizes the depth code command and automatically adjusts the wings to the required tilt. The birds are powered by rechargeable batteries to provide system operation for a number of weeks.

Remote depth control has the following advantages:

1)   The cable does not have to be reeled in to change the depth settings. (This convenience is the biggest advantage).

2)   The cable can be submerged to a safe depth if a vessel should attempt to cross it.

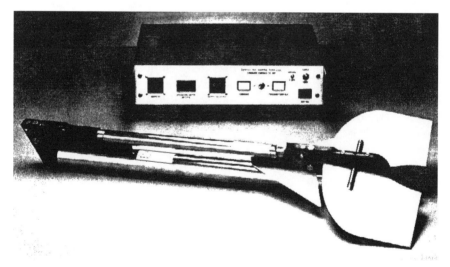

Fig. 54. Active depth controller "bird" (courtesy Syntron, Inc.).

3) It can be raised to the surface to avoid reefs or shoals.
4) Lines crossing from deep water into shallow water can be shot at optimum cable depth (but with inherent change in the receiver ghost effect).

2.2.4 Streamer Depth Indicators

The streamer depth as well as lateral position can be monitored from the instrument room using remote sensors. The depth is usually measured at a number of locations along the cable. The sensor placed in the streamer is a coil whose electrical inductance changes as a function of pressure. The coil is connected through a pair of wires along the streamer to the depth indicator system in the instrument room. With a calibration of depth to pressure to coil inductance, the visual readout provides a continuous indication of cable depth with an accuracy of ±5%. The depth sensors or *depth sections* usually are placed at the same point as the depth control birds to monitor cable depth and bird performance. This is probably satisfactory with a well-ballasted cable but could be deceptive if the cable is not neutrally buoyant.

2.2.5 Streamer Heading

Streamer compasses can be placed in or on the cable to detect heading. With $1°$ accuracy, 10 or more are often used to detect cable drift (i.e., feathering) off line. Ships have computer systems that continuously log the compass data and process them to produce an on-board plan view of the streamer position with respect to the preplotted desired line to be shot. This allows the crew to adhere to streamer feathering specifications as well as allow the computation and storage of common midpoint (CMP) locations needed during 3-D surveying (see Chapter 7).

2.2.6 Streamer Noise

Good design of acceleration-canceling hydrophones has limited much of the noise to the extent that in a calm-sea state, signal-to-noise ratio frequently is better in marine data than in land data. Noise is also generated by vessel mechanical vibration, cable strumming, and the vortex action of water around the cable. All such actions result in forces that cause unwanted noise. In the 1940s and 1950s, it was customary to stop the boat dead to minimize these noise sources during recording operations. A continuous tow cable was then designed to suppress such acceleration noise by combining the hydrophones as explained earlier.

The following noise types can occur in a modern cable when cables are towed through water:

1) *Depth controller noise*—Water flow turbulence along the streamer and
    over the birds may be reduced if birds are placed away from live
    hydrophones. Such turbulence also can generate extreme noise when
    birds are diving hard.
2) *Poor-ballast noise*—If a streamer is poorly ballasted, the birds will tilt
    their wings to greater angles trying to pull the streamer up or down to
    the desired depth. As a result, local turbulence is generated which pro-
    duces *bird noise* on the streamer.
3) *Rough sea state noise*—Sea swell causes up-welling and down-drafting
    of volumes of sea water. This turbulence often generates a short-wave-
    length vertical pressure wavefield causing individual live sections to
    raise up or drop down depending on the direction of the surface swell.
    When individual live sections are moved relative to adjacent live sec-
    tions, high-amplitude noise bursts are observed. In rough seas, as a
    rule of thumb, a maximum depth of turbulence which causes unac-
    ceptable streamer noise bursts occurs at half a wavelength of the swell
    beneath the swell trough (Figure 55). Such noise bursts can be over-
    come by trough shooting (Figure 56). During trough shooting (a tech-
    nique often used in the North Sea where the weather window is
    limited), the troughs drop the whole cable down, and it is lifted up

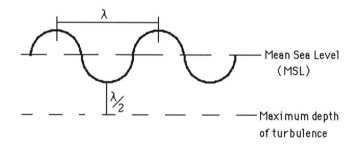

Fig. 55. Rough sea turbulence.

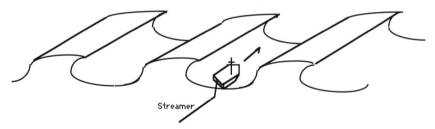

Fig. 56. Trough shooting.

equally when the next wave peak arrives. Consequently, if lines can be shot at right angles to the sea swell direction, production can continue with acceptable noise limits. One problem associated with this type of recording is that individual shots are recorded with the streamer at different depths to the seabed, which may deteriorate stack quality. It is therefore not recommended but is often the only alternative to recording no data at all. Also, in shallow-water areas where streamer depth must be reduced, surf and waves breaking at the sea surface can contribute significantly to background coherent noise. This is a constant problem with shallow water and transition-zone recording, about which nothing can be done.

4) *Ship propeller noise*—Propeller noise may travel horizontally through the water column and be recorded by the cable. Alternatively, if a shallow hard seabed exists, a propeller beat frequency can be set up by the pressure wave from the propeller being reflected back to the surface by the reflective seabed. This noise can be attenuated during data processing by trace mixing and stacking. Seismic vessels either have a single propeller (in which the speed is adjusted by the blade pitch) or twin propellers (in which the speed is adjusted by changing blade rotation velocity). If the vessel has twin propellers, the individual propeller speeds may be adjusted so that the beat frequency is reduced. With a single-propeller ship, changing the pitch may not resolve the problem and, as a last resort, the vessel may have to be changed.

2.3 Fundamentals of Array Design

Noise is a major problem experienced in seismic recording. If noise has a consistent, repeatable form, we should be able to recognize it and, hence, attempt to attenuate it. Once recognized, repeatable noise can be canceled using arrays of phones. First we must understand the types of noise with which we have to deal.

As was seen in Chapter 1, ground roll (a Rayleigh or surface wave) is a form of noise that occurs when the weathering layer is excited by a shot being fired. Because ground roll has particle motion that is circular to elliptical, the noise has a vertical component that shakes the phones vertically as the noise travels along the surface. On a shot record, ground roll appears as lines of coherent noise overriding the data; that is, the ground roll has a recognizable pattern from one trace to the next. In Figure 12, coherent noise appears on the seismic shot record as event a, masking the reflected event c at the shorter offsets. Noise that does not occur in repeatable patterns is called *incoherent* or *random noise*; area D in Figure 5 shows an example. Random noise can be removed by stacking, so usually the greatest concern in the field is how to remove coherent noise from the field records.

Noise attenuation in the field is preferable because the high amplitude of coherent noise may not allow low-amplitude reflection signals to be seen. Coherent noise is continuous across the seismic record. Random noise does not have this property but may be simply observed on the seismic record as occasional high-amplitude noise bursts. Halfway between these is pseudo-random noise in which receivers are spaced so far apart that the coherent noise train does not appear linear across the record.

Random noise may be attenuated simply by spacing the elements of a receiver array far enough apart to ensure no correlation between the noise detected on those elements. For example, a gust of wind may affect two geophones similarly if they are positioned close to each other, but the effect becomes random if they are spaced so far apart that only one of the geophones is affected. When elements are spaced far enough apart for such attenuation, the noise reduction factor is $n^{1/2}$, where n is the number of geophones. Therefore element spacing is a factor to be considered.

For coherent noise attenuation, size, spacing, weighting, and orientation of the array also must be considered. With ground roll, noise trains travel horizontally from the shot with distinct velocity, wavelength, and frequency characteristics. As ground roll travels along a horizontal surface, the particle motion is roughly elliptical in the vertical plane containing the direction of propagation. As a consequence, ground roll vibrates geophone elements vertically. The geophones can be arranged so that their summed signal will cancel the vertical component of the horizontally traveling noise trains while adding the vertical component of the reflections.

Consider a simple horizontal wave (Figure 57), with two receivers close together. As the horizontal wave moves past the two receivers, provided they are spaced a half wavelength apart, their total vertical output will sum to

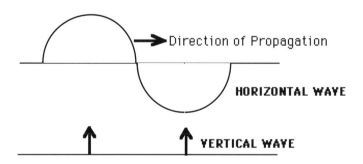

Fig. 57. Horizontal and vertical waves.

zero. However, the arrival of a vertically reflected wave will cause an additive output from both geophones since the wavefront is *in phase* at the receiver array. This is how we can attenuate or *suppress* the horizontal wave while enhancing the useful vertical signal wave.

In shallow-water areas, an equivalent ground-roll type wave exists that correctly is called the Scholte wave. This is referred to in the industry as bottom roll or mud roll, which the geophysicist should expect when designing shallow-water marine surveys. Here, we will use land-noise problems to illustrate the concept of receiver array design. These concepts apply equally well to land and marine source array design.

2.3.1 Synthetic Record Analysis

A *synthetic record* is a record numerically constructed from a model of the earth, as opposed to a *field shot record*, which is a measurement of the earth's response to a shot. We will analyze a simple synthetic record to illustrate the concepts of noise attenuation. Consider the schematic shot record of Figure 58. This involves a simplified five-channel (or "station") land record. The shot has been fired 25 m from the nearest geophone, with all geophones spaced 25 m apart. Events A and B are seen at each geophone station. Event A takes a shorter time to cross the stations, whereas event B takes longer.

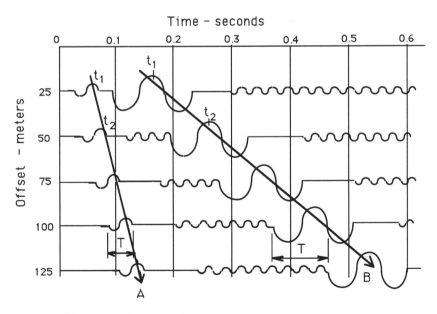

Fig. 58. Schematic shot record.

Obviously, A must be a wave moving faster than B. Note also that wave B takes longer than wave A to complete one full cycle. The length of one cycle of a wave is called its *periodic time*, *T*, and is inversely proportional to a wave's dominant frequency. That is,

$$f = \frac{1}{T}.$$

(26)

The relationship between velocity (*V*), wavelength (*λ*), and frequency (*f*) of a wave, is given by

$$V = f\lambda.$$

(27)

The velocity of a sound wave is a vector quantity that cannot be measured unambiguously by a single line of receivers. For example, if a wave pulse appears simultaneously at all geophones in a linear spread, then we could interpret that as either 1) a wave traveling perpendicularly to the spread (with unknown speed), or 2) a wave traveling parallel to the spread (with an infinite speed). To apply the above equation to ground roll, *V* is interpreted as the apparent horizontal velocity. The true velocity can be determined only if the wave's direction of propagation is known.

In the record in Figure 58, we can determine the ground roll's apparent velocity simply by working out how fast the same point (say a peak) on the wavefront moves past the detectors (25 meters apart). The periodic time *T* is determined by finding the time between two adjacent trace peaks of the same arrival, and the dominant frequency can then be calculated. Hence, since wavelength *λ* = *V*/*f*, the ground-roll wavelength may be computed. Let us now show in detail how the above ideas can be used to design a simple, two-element array that will attenuate event B in Figure 58.

1) Select a group of traces that has a representative sample of the wave to be attenuated. Take any two adjacent peaks or troughs of the wave where the wave has similar character (or shape) to that on the next adjacent trace. For example, the largest amplitude wave on the second trace (at 50 m offset) has troughs at about 0.22 and 0.31 s. The time difference is therefore 0.09 s, and since *f* = 1/*T*, the dominant frequency is about 11 Hz.

2) The strong, low-frequency (11 Hz), high-amplitude wave going across the synthetic record is now inspected to find its apparent velocity. The peak of the low-frequency wave appears on trace 1 at t_1, which is at about 0.17 s (at 25 m offset from the shot) and on trace 2 at t_2, at about 0.26 s (at 50 m offset from the shot). So, the wave has traveled 50 m –

25 m = 25 m in 0.26 s – 0.17 s = 0.09 s. That is, the wave has moved at an apparent velocity of 25 m per 0.09 s or 278 m/s.
3) The wavelength is V/f or $278/11 = 25$ m.
4) Any wave is best canceled by adding the maximum positive values to the maximum negative values; for a perfect sinusoidal wave, the result is zero. So, to position two geophones to cancel such a wave, they must be a half wavelength apart. In this case, the geophones must be 25/2 apart or about 12 m apart.

In actual field data, ground roll is comprised of many wavelengths and, furthermore, its composition may change from point to point along a line. For this reason, many elements are used in the array instead of just two. Some arrays may even be spread in a 2-D pattern to eliminate broadside noise as well as in-line noise.

The ground roll's bandwidth, as a rule of thumb, is considered to extend from 0.5 to 1.5 times the dominant frequency calculated. Hence, ground roll with a dominant frequency of 20 Hz often has a bandwidth of 10–30 Hz, while a dominant frequency of 30 Hz has a bandwidth of 15–45 Hz.

2.3.2 Receiver Array Design

As the previous example showed, receivers may be positioned to cancel particular waves. So far, the use of just two receivers has been considered. One way to expand this number is to space the geophones equally along a line. This pattern of phones is called a *uniform* or *linear* array. Variations of this, such as *weighting* of the array, are possible.

Before considering array design, it should be understood that, in practice, field arrays do not have ideal characteristics. Arrays may fail to perform optimum attenuation of ground roll because of

1) Erratic element spacing because of terrain or poor placement (land)
2) Elevation variation (land) or cable depth variation (marine)
3) Arrival time variations because of weathering thickness variations
4) Detector sensitivity and coupling variations
5) Change in surface-wave characteristics because of changes in weathering-layer lithology parameters

However, even if arrays reduce coherent noise only marginally, their use is still an important step because a minor reduction in ground roll is an increase in data quality and an improvement in signal-to-noise ratio.

2.3.2.1   The Uniform Array

Because sound waves can approach an array of receivers from any direction, characterizing an array's response is, in general, a 3-D problem. Here, however, we will consider only a much simpler, 1-D problem. The full 3-D theory is described in Appendix B.

Suppose a set of geophones are laid out in a straight line on the surface and we are interested in how that array responds to a sound wave that travels horizontally and parallel to the array. A convenient way to characterize such a wave is by its *horizontal wavenumber, k,* which is the reciprocal of the wave's wavelength. Thus, we have the following relationships:

$$\lambda = 1/k = V/f. \tag{28}$$

Note that $k = 0$ implies that the wavelength is infinite. This can occur either because a wave's velocity is infinite or because the wave's frequency is zero.

Before explaining how to calculate an array response for the simple 1-D problem, it is worth noting that the method can have some uses even for sound waves propagating in two dimensions. Figure 59 shows some waves whose raypaths lie in the vertical plane below the 1-D linear array. Clearly, the propagation of such waves has two components, one vertical and one horizontal. The 1-D array response method applies to the horizontal component. Often geophysicists speak of the horizontal component by using the word "apparent." Thus, the *apparent velocity* or *apparent wavelength* of a sound wave refers to the velocity and wavelength of the horizontal component. A wave that travels vertically has a zero horizontal wavenumber and an infinite apparent horizontal velocity. This is just another way of saying that a vertical wavefront is detected simultaneously by all the surface geophones.

Figure 60 shows the response R of a five-element horizontal linear array having uniform element spacing d. The figure shows cancelation of waves of six different wavelengths, ranging from λ = infinity to λ = 0.8. Numbers on the right show the product of phone spacing d with wavenumber k, assuming a phone spacing of unit value 1. If a wavelength λ is infinitely long ($k = 0$), all phones observe equal amplitude and the total output response R is 5. If the

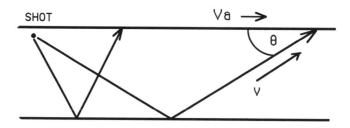

Fig. 59. Wavefront raypaths arrive at points along line.

wavelength is 4 units, the response is –1, and so on. A graph of the responses in Figure 60, using a linear output response scale R versus a linear wavenumber k scale for unit geophone separation d, is shown in Figure 61.

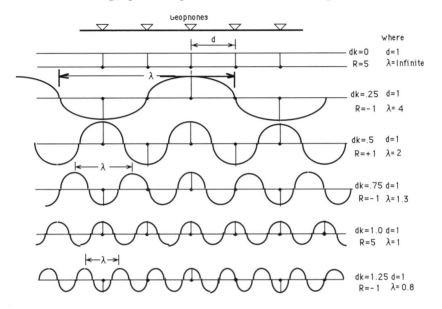

Fig. 60. Geophone array response.

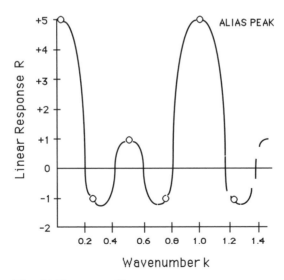

Fig. 61. Response/k curve.

If we continue shortening the wavelength, a point is reached where the wavelength is exactly one phone spacing. The response from this point on is identical to that observed over the 0.0–1.0 wavenumber range. We have reached the alias or *fold-over point*, where even shorter wavelengths can appear as the longer wavelengths (see "Digital Sampling," Chapter 4). This unwanted situation means the array has passed the point where it can provide maximum wavelength attenuation. In Figure 61, this occurs in the region of $k = 1.0$, and hence wavenumber values in excess of 1 just repeat the response of the array.

For the theoretical approach to the computation of array response, refer to Appendix B. From that theory, and using the decibel form of notation,

$$\text{Amplitude response (dB)} = 20 \log \frac{\sin(\pi dkN)}{N \sin(\pi dk)}, \qquad (29)$$

where d is the element spacing and N is the number of elements. The amplitude response of a six-geophone array with 5-m phone spacing is shown in Figure 62.

Attenuation is maximum at five notch points occurring at $1/Nd$, $2/Nd$,..... $(1-N)/Nd$. The first alias peak, at wavenumber $1/d$, has no attenuation. A feature of a uniform array's attenuation characteristics is that the attenuation *lobes* form a curve or "envelope" of attenuation, where the amplitude of the array response at the envelope base is $1/N$, or $20 \log 1/N$ (dB scale).

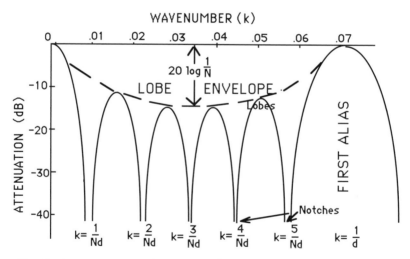

Fig. 62. Array response for six geophones 5 m apart to horizontally traveling waves.

Because of the $1/N$ factor, overall attenuation increases as array length increases, or, viewed alternatively, as the number of phones increases. The center of the lobe envelope is at wavenumber $1/2d$. Thus, overall array length ideally should be at least twice the longest wavelength to be attenuated.

A very long array has good attenuation of horizontal noise, but it also attenuates high-frequency reflections. Shallow reflections, which have undergone a minimum of absorption attenuation because of their short travel-path length, arrive at shallowest incident angle θ. Because of the small incident angle, shallow reflections can have a significant component of motion in the horizontal direction (see Figure 63) and are therefore affected by array attenuation. In comparison, deep reflections undergo greater absorption attenuation because of their longer travel-path length, but deeper reflections arrive at steeper incident angles and are not as affected by array attenuation. Consequently, shallow, high-frequency reflections may be attenuated by an array to a greater degree than the deeper, low-frequency reflections.

This is therefore a good reason to maintain a short array length, which must be designed for the retention of the target (shallow or deep) reflection. The use of field arrays always entails a compromise. Long arrays can achieve better noise attenuation than short arrays, but at the expense of partial attenuation of the desired signal.

2.3.2.2   The Weighted Array

Coherent noise trains often consist of a number of different wavelength noise trains that gradually separate with distance along the receiver line, as they are traveling at different velocities. The shot record of Figure 3 shows a good example of this phenomenon, which is referred to as a *dispersive wave* because the initial fundamental ground-roll wave at short offset separates out into packages of energy or waves. That is, the noise disperses into separate wavetrains or higher-order-mode waves. If a number of different noise

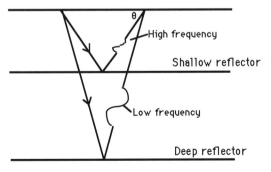

Fig. 63. Shallow and deep raypaths.

wavetrains, each having different values of k, are to be attenuated, it would be useful to have an array that could be designed to attenuate those specific values of k. The weighted array provides this option.

The problem of designing weighted arrays and computing their response can be approached by thinking of weighted arrays as composites of simpler unweighted subarrays. Consider the geophone array depicted below in which small circles represent phones as if the array were laid out on the ground and the two phones at the center of the four-phone array show that the array is weighted at the center by a factor of 2:

$$o$$
$$o \blacktriangleleft\!\!-d\!\!-\!\blacktriangleright \quad \blacktriangleleft\!\!-d\!\!-\!\blacktriangleright o$$
$$o$$

This figure is schematic, representing a weighted array, and is not how the array is laid out on the ground (imagine the center phone gives an output twice the amplitude of the other phones, so we represent it by two phones). The array therefore has a weighting of (1,2,1), where the number of phones at each receiver point is represented by the numbers in parenthesis. This array is just the same as the composite of two simpler unweighted identical subarrays depicted below.

First subarray
$$o \blacktriangleleft\!\!-d\!\!-\!\blacktriangleright o$$

$$o \blacktriangleleft\!\!-d\!\!-\!\blacktriangleright o$$
Second subarray

For a linear unweighted array, the maximum ground-roll cancelation occurs when $\lambda = 2d$. Thus, in this simple case, we can consider the weighted array as a single linear subarray repeated again after lining up first and last phones. The response of the resulting weighted array can then be obtained by convolving the response of the simple basic subarray with a repeat factor representing the number of times the original basic subarray has been repeated.

The mathematical process of convolution is shown diagramatically below, where the weighted array on the left (1,2,1) is the same as the two subarrays on the right (1,1,0) and (0,1,1), where the "0" shows no phone present in that position, and the "1" indicates that there is a phone present in that position. As above, circles represent phones.

			1st	2nd	3rd Postition
o	o	o	o	o	
	o			o	o
1	2	1	1	2	1

In the equivalent array on the right, the first two-phone subarray has one phone in the first position and one phone in the second position. Hence, that subarray is represented by (1,1). The second subarray is identical but has been shifted along one position. This shift is the *repeat factor*, which can be represented by (1,1), where the first "1" shows us that basic subarray started in the first position (denoted by 1) and the array then moved to the second position (denoted by 1). So, the repeat factor (1,1) shows that the basic subarray moved one position. Note that (1,1) is the same as (1,1,0), since the 0 only shows that either there is no third geophone (where it represents a geophone array) or the basic array is not repeated (where it represents the repeat factor).

Convolving the basic array with the repeat factor produces the same representation as the weighted array (which was 1,2,1). A convolution is performed by reversing the order of the first numbers and shifting them across the second number series while multiplying. We use a star (*) to represent this process of convolution. In our example, we have to show that the convolution of (1, 1) * (1, 1) = (1,2,1).

For example, arrange the two arrays by reversing the first array numbers (where we represent 1,1 by 1,1,0) to become (0,1,1) and write the two arrays in two columns. The numbers at positions A and B will be the first to be multiplied.

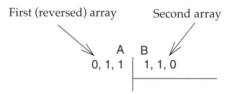

1) Now multiply 1 at A by 1 at B by placing the 1 at A underneath position B (arrowed) and position the resulting value of 1 at C.

2) Now we shift the first array numbers along and multiply again. To do this, we put the last two numbers of the first array under the first two numbers of the second array and multiply them so that the result can be written down in position D.

$$
\begin{array}{c|l}
 & \text{A} \quad \text{B} \\
0, 1, 1 & 1, 1, 0 \\
 & 1, 1 \\
\hline
 & 1 \\
 & 1 \ 1 \ \text{D}
\end{array}
$$

3) Do the next shift of the first array, multiply with the second array, and put the result at position E.

$$
\begin{array}{c|l}
0, 1, 1 & 1, 1, 0 \\
 & 0, 1, 1 \\
\hline
 & 1 \\
 & 1 \ 1 \\
 & 0, 1, 0 \ \text{E}
\end{array}
$$

4) Perform the next shift of the first array and multiply again, putting the result at F. Note that the last 1 of the first array is not written down (at G) because there is nothing to multiply it by.

$$
\begin{array}{c|l}
0, 1, 1 & 1, 1, 0 \\
 & \quad 0, 1 \ \text{G} \\
\hline
 & 1 \\
 & 1 \ 1 \\
 & 0 \ 1 \ 0 \\
 & \quad 0 \ 0 \ \text{F}
\end{array}
$$

5) The final shift puts 0 under 0 at the G location so the multiplication contributes nothing to the exercise. Finally then, we add up the rows of numbers to produce a resultant answer for the convolution of the two arrays. The result is shown in the left column as 1,2,1.

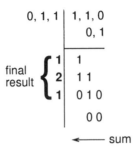

This is known as the *convolution table* and is the final result of the convolution of the subarray and its repeat factor. Thus, the convolution of subarray (1,1) with repeat factor (1,1) gives the array (1,2,1). So far, we have convolved two very simple uniform arrays to obtain the response of a simple weighted array. Here is a slightly more complicated example.

Consider the weighted array (1,2,2,1), which appears below using our earlier notation.

$$
\begin{array}{ccc}
\circ & \circ & \circ \\
& \circ & \circ & \circ \\
\end{array}
$$

This pattern is a simple three-element uniform subarray that appears twice. As in the previous example, the repeat factor is (1,1). Thus, the weighted array response is given by (1,1,1) convolved with (1,1).

The first 1 in the repeat factor represents the first subarray, while the second 1 represents the number of times that subarray has been repeated at the second position. Thus, the repeat factor establishes the position of the second identical array with respect to the first.

Proceeding as explained above, the convolution table is as shown below.

$$
\begin{array}{c|c}
1 & 1 \\
2 & 1\ 1 \\
2 & 1\ 1 \\
1 & 1 \\
\end{array}
$$

Sum ◄——

Thus, $(1,1,1) * (1,1) = (1,2,2,1)$. This procedure works with any number of subarrays, provided the repeating subarrays are identical.

We now have a simple mathematical tool that allows us to determine the response of complex weighted arrays. The response of any complex arrangement of identical simple arrays can be described as a convolution of the simple array pattern with a repeat factor. The amplitude response formula (1) of Appendix B is then applied so that lobe and notch values can be computed for the array and its repeat factor on the same graph, then summed. The resulting response of the simple array plus its repeat factor is the total response of the complex array. The simplicity of this treatment allows the field hand to compute response curves rapidly by hand, without field computers.

(Remember that summing two negative dB values produces a larger negative dB value. When we refer to the attenuation of waves, a 0 dB value is a wave's peak value, and attenuation of a wave reduces values. So, reducing values from a peak in dB terms is the same as subtracting dB numbers to become more negative. Further attenuation of a wave increases the –dB value. For example, –20 dB plus –10 dB produces an attenuation of –30 dB.)

An array pattern that repeats after two shifts rather than one can be accommodated using the repeat factor $(1,0,1)$, where the zero shows that the shift did not stop at the second point but continued on.

$$O \qquad O \qquad O \qquad O$$
$$O \qquad O \qquad O \qquad O$$
$$(1, 1, 1, 1) * (1, 0, 1) = (1,1,2,2,1,1)$$

A uniform array with equal phone spacing but consisting of odd numbers which are some function of the full station length also can be accommodated. Consider a nine-phone array in this pattern where the station length is, say, 8/3 units. Each phone may be 1/3 units apart, so that it looks as shown below.

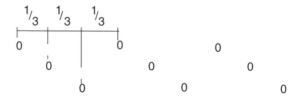

This can be considered as the basic array along the top line which would be $(1,0,0,1,0,0,1)$ convolved with the repeat factor $(1,1,1)$ where the distance apart $d = 1/3$. It also will appear as a nine-element array to waves traveling from left to right because the array is unweighted and phones are of equal separation distance (1/3 between phones).

Therefore, the response will be identical to a nine-element uniform array.

2.3.2.3 The 'Superposition' Pitfall

One mistake to avoid is attempting to compute the response of a complicated array by summing the responses of simpler arrays. This is inherently incorrect because it does not account for the change in phase as a result of the horizontal wave arrival. The phase change is accounted for correctly in the convolution method repeat factor.

For example, consider a simple array of two phones with in-line separation d. This would look as follows.

o

   o

The incorrect approach would say this is a single phone array------   o
and added to this is the other single phone array----------------------   o.

Since the response of a single element is flat, the sum of the two single-element responses is also flat. Obviously, this is an incorrect result, which illustrates the pitfall of just summing responses.

Land geophone arrays frequently use weighted patterns. An alternative approach is to position phones (and hence stations) very close together in linear-array form and record a very large number of stations. This allows array weighting to be performed in the processing center by mixing data recorded by the closely spaced stations. However, the choice is not always which approach is better for signal-to-noise ratio improvement but which approach is more economical.

The convolution approach also may be used for computing the response of an areal 2-D pattern.

Occasionally, another type of array is used, called the cosine or Chebychev array. This type of array has weighting like a cosine curve peaking at the center, the response of which can be computed by the convolution theory. Arrays also can be simulated in a computer to produce a desired response. For example, if a number of very short linear arrays are used in the field (on land, geophones may be placed close together; in marine recording, short streamer groups may be used), we can sum them to produce a single longer array (this is referred to as *mixing* the receiver stations) which will cancel long-wavelength ground roll. This initial computer processing step may be performed in the field or in the processing center. If a known array response is therefore required, receiver station mixing can be a useful step, provided the receiver interval is short enough to allow it.

Another array approach, called the *stack array*, is an extension of the application of data processing to remove ground roll. Developed by Anstey (1986a), it basically builds long arrays using a particular source/receiver geometry to remove ground roll by the stacking process. An acquisition geometry known as *symmetric sampling* (see later this chapter) adds to this by using a source array, which helps increase attenuation of ground roll. These will now be discussed in detail.

2.3.2.4   The Stack Array

As discussed in Chapter 1, the signal-to-noise ratio is improved by stacking the data. For some years, many geophysicists often preferred to lay out uniform geophone arrays and simulate weighted arrays by trace mixing individual receiver station data in the processing center to remove ground roll. Their reasoning was that this approach was often less expensive and cumbersome than laying weighted geophone arrays, and frequently appeared to do just as good a job. Anstey (1986a) was the first to explain the obvious. If geophones are equally spaced along the full length of each station, and the station interval is short enough for correct subsurface sampling, then the traces in a CMP gather are distributed evenly and continuously across the gather. By the application of NMO and subsequent stacking of these traces in data processing, the ground roll observed on shot records is canceled. Anstey explained the logic behind what had been practiced in the processing center for years and called this the *stack-array technique*. He also went on to suggest recording configurations that would help the stack array to further suppress ground roll.

Ground roll is a function of space and time. The wave moves along the surface (and across the geophone arrays) horizontally and down the traces in time. Using the synthetic time-space (*t-x*) shot record of Figure 64, imagine a series of panels of seismic traces with the ground roll or *noise cone* passing diagonally across the record. In the top diagram of Figure 64, the seismic trace panels are separated by the six straight lines with the last synthetic trace displaying the ground roll as it might normally be seen on the record. The geophone groups producing these panels have a linear array with equally spaced geophones covering the whole spread of stations along the line.

The bottom diagram of Figure 64 shows a series of individual traces that will be resorted to make up panels of CMP traces. The ground roll passes across the individual traces, and if the appropriate traces had NMO correc-

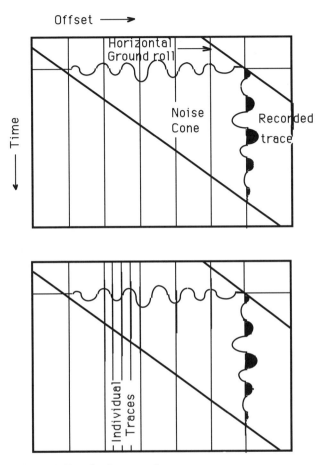

Fig. 64. Synthetic record.

tions and then were added together or stacked to produce one CMP trace, the result would be the same as having a longer geophone array to cancel the ground roll wavelength, as we did earlier using linear geophone arrays. This is stacking in its simplest form. The result is cancelation of ground roll to a degree dependent on the ground-roll complexity. This is displayed synthetically for each panel in Figure 65.

Anstey pointed out that particular arrays used in the field would assist the breakup of ground roll in a CMP gather and called the use of these arrays with stacking the *stack-array criterion*. Array geometries could be determined prior to the commencement of a survey, so that there would be no need to perform noise tests. Anstey described a number of source-receiver configurations that would satisfy the stack-array criterion. If trace mixing is employed in data processing, then the field layouts would be as shown in Figure 66. The trace mixing allows stations to be shorter, as the mix doubles the effective length of the station. One problem with such mixing is that the high frequencies at steep reflection dips are destroyed (Anstey, 1986b).

In Figure 66, the flags indicate the centerpoint of the receiver stations. The black dot indicates the shot location. Where the spread is split-spread, the

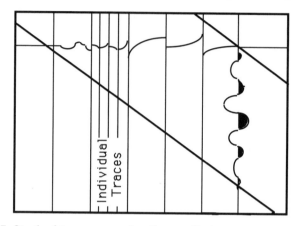

Fig. 65. Stacked trace ground roll cancellation in a CMP gather.

Type	Spread	Field layout	Trace mix	receiver interval	receiver length	source interval	CMP interval	Channels	Fold
1.	Split	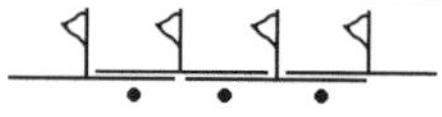	2:1	35	70	35	17.5	120	60
2.	Split	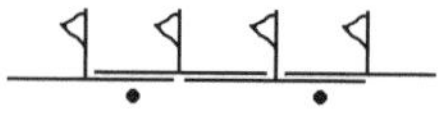	2:1	35	70	70	17.5	120	30
3.	Split				NOT RECOMMENDED				
4.	Off-end	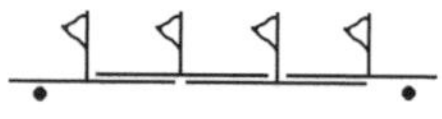	2:1	35	70	35	17.5	60	30
5.	Split	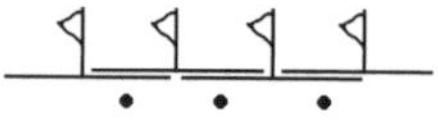	2:1	17.5	35	17.5	8.75	240	2x60
6.	Split	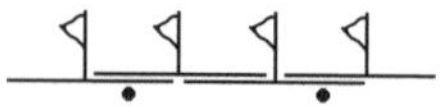	2:1	17.5	35	35	8.75	240	60
7.	Off-end	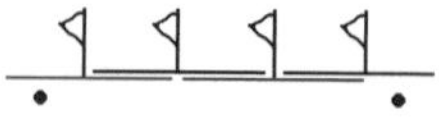	2:1	17.5	35	17.5	8.75	120	60

Fig. 66. Stack-array configurations.

shot location is shown in its correct location for continuous coverage. Where the spread is off end, the shot is shown at each end to emphasize an off-end configuration, but shot locations are still at the value stated in the source interval column where distances are in meters. The split-spread configuration (Type 3) with the shot at the station is not recommended.

The stack-array criterion therefore requires an even, continuous, uniform succession of phones across the CMP gather, so that:

1) In marine operations using an off-end spread, the group length must be equal to the group interval, and the source interval must be half the group interval.

2) In land operations, split-spread recording ideally should be used in which the station length is equal to the station interval, the source interval equal to the station interval, and the shot is between the stations.

For split-spread recording, shooting between the stations is preferred because this provides more continuous array coverage and hence a better chance of ground-roll suppression.

The common-shot CMP stacking diagrams described in Chapter 1 can explain how ground roll can be attenuated. Two alternative ways to represent stacking diagrams are shown in Figure 67.

Both diagrams represent six adjacent shots into a split-spread array. On the right, the CMP coverage is shown in the customary manner with midpoints shown horizontally. The gap is the split-spread configuration gap. On the left, the midpoints are diagonal with one half of the spread above the line and the other half below the line. The fold is four at the chosen CMP. The requirement for continuous receiver coverage is shown for both methods in Figure 68.

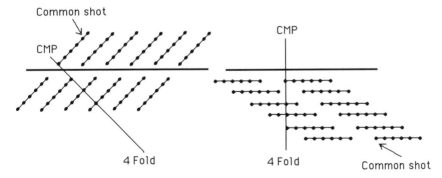

Fig. 67. Stacking diagram styles.

Anstey uses the method on the left to demonstrate continuous coverage, in which the bottom half coverage lines (e.g., lines 1 and 2) must fit like a comb into the top half to produce continuous coverage. Alternatively, with the method on the right, the stacking diagram left half must show lines of coverage that fit into and between the lines of the right half (a line of coverage is needed where the dotted line is shown) to provide continuous coverage.

An alternative approach to understanding the benefits of shooting between the stations is to consider near-offset travel paths. The signal-to-noise ratio in CMP data is increased if different raypaths are stacked to make the CMP trace. Using split-spread geometry with the shot fired at the receiver station, and looking at the near-offset traces with the conventional method of drawing stacking diagrams (the one on the left in Figure 68), we can draw the raypaths as the spread rolls along the line. In Figure 69, the raypaths from SP1 are shown as expected. When SP2 is fired, the raypath to the right of SP1 is duplicated by the raypath to the left of SP2. Instead of the CMP between SP1

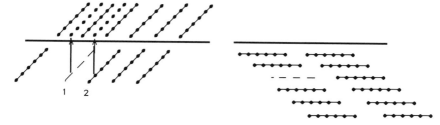

Fig. 68. Stack-array coverage.

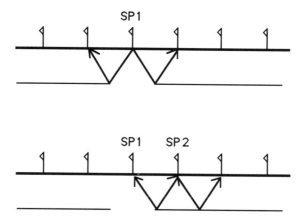

Fig. 69. Split-spread shot on station.

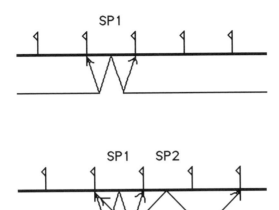

Fig. 70. Firing between the stations.

and SP2 being two horizontally stacked traces, it becomes two vertically stacked traces (i.e., instead of two-fold, it is two times single fold).

If the shot is fired between the receiver stations, SP2 has different ray-paths, allowing correct CMP stacking, as shown in Figure 70. Consequently, the simple rule is that when shooting off-end, fire the shot at and between the stations, whereas when recording split-spread, fire the shot between the stations.

2.4 Symmetric and Asymmetric Recording

In symmetric recording, both source arrays (like vibrator arrays) and receiver arrays are used. In asymmetric recording, the source is a point source (such as a single-charge explosive) and the receiver has an array. In updip or downdip recording, symmetric recording is preferred because this will avoid the static effects caused by using only a receiver array (Vermeer, 1991). For example, in Figure 71, an updip shot from a single energy source into a receiver has the reflections arriving nearly vertically and at similar times. However, the single shot downdip has the reflectors arriving at an acute angle, and at different times. The result is a different stack for the updip versus downdip recording. If split-spread recording is used, then the dipping events from updip recording will not stack in well with those from the downdip recording and reflections may not image well.

By comparison, the use of symmetric arrays provides similar travel paths (though not identical since there are normally more receivers than sources), and hence Figure 71 makes the point that we always should use symmetric

Asymmetric Configuration

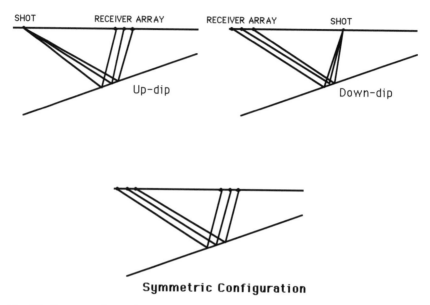

Symmetric Configuration

Fig. 71. Symmetric and asymmetric recording.

arrays where possible. The choice of the array parameters is dependent on noise tests performed in the area of operations.

2.5 Receiver Ghosting

When reflected seismic waves travel vertically upward toward the receiver array, some of the energy can become trapped within the weathering layer by rebounding off the air-ground interface (and air-sea interface in water), internally reflecting down at different transmission angles. Such waves again may reflect upward from the weathering base, causing what is known as a *reverberation*. Such a reverberating wave is unwanted and is considered as a *ghost* of the earlier-arriving wave. Techniques to attenuate such waves in the field are discussed later in Section 3.5. (With land receivers, there is little that can be done to attenuate such waves apart from changing the source depth). Marine techniques using dual sensors were discussed in Section 2.1.4.

Exercise 2.1

Find the wavelengths of the three waves indicated in Figure 72. Trace spacing (geophone interval) is 5 m. Seventy-two traces have been recorded in three groups of 24 traces.

Exercise 2.2

On a graph of attenuation (0–40 dB) versus wave number (0–0.2), sketch the response of a six-element array with 5-m uniform spacing.

a) Note notch points, lobes, and first alias peak.
b) Calculate attenuation at the lobes.
c) Calculate maximum lobe envelope value.
d) What is the effect of reducing the interval to 3 m?

Formulae, Response (dB) $= 20 \log \dfrac{Sin \pi dkN}{N Sin \pi dk}$

$$k = \frac{1}{Nd}, \frac{2}{Nd} \ldots \text{(notches)}$$

$$= \frac{1.5}{Nd}, \frac{2.5}{Nd} \ldots \text{(peaks of lobes)}$$

(Answers: Lobes are at 0.0495, 0.0825, 0.115, 0.149. Notches are at 0.033, 0.066, 0.099, 0.132, 0.165. Attenuation is 12.55, 15.26, 15.26, 12.55. Maximum lobe envelope value is 15.56 dB.)

Exercise 2.3

Show that a weighted array (1,2,3,3,2,1) can be modeled as three uniform subarrays (1,1,1,1) with a repeat factor of (1,1,1).
The response of such an array is shown in Exercise 2.4.

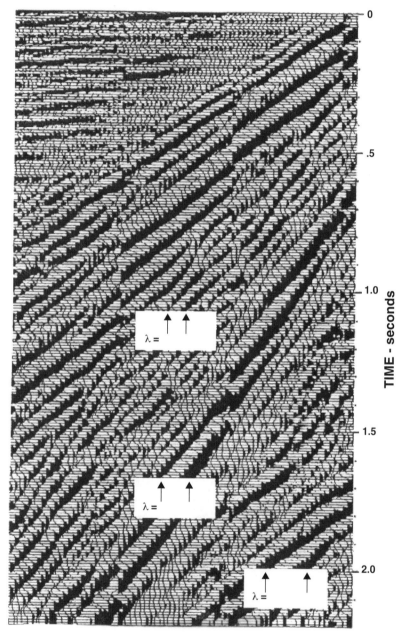

Fig. 72. Wavelength measurement.

Exercise 2.4

Determine the geophone separation for an array with weighting (1, 2, 3, 3, 2, 1) for the response curve shown in Figure 73.

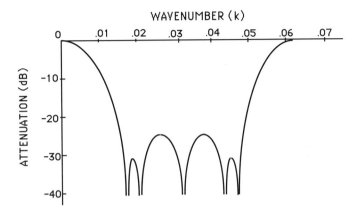

Fig. 73. Response curve.

Chapter 3

Seismic Energy Sources

For many years a single explosive charge was the most often used source of seismic energy. A single charge is an *impulsive point source;* that is, all of its energy is generated at one time in one location. For small charges, the amount of seismic energy produced per shot can be increased simply by increasing the charge size. However, as years of experimentation have shown, there are diminishing returns of seismic energy as large charges are made even larger. Single impulsive point sources cannot efficiently produce the amount of seismic energy needed to image deeper targets well. Three strategies evolved for overcoming the limitations of an impulsive point source: (1) distribute the source energy in space, (2) distribute the source energy in time, and (3) distribute the source energy in space and time.

A source can be distributed in space by dividing the single large charge into smaller point charges and firing them together in spatial patterns. The resulting *source array* produces more seismic energy than does its single-charge counterpart. As an additional benefit, the array pattern can be designed to reduce noise problems, just as can be done with a receiver array (see Chapter 2). Primacord is an example of a spatially distributed explosive source.

There are two ways a source can be distributed in time. In the discrete method, a single massive charge is replaced by many small charges that are fired sequentially from a single shotpoint. The resulting data records are stacked to simulate the shot of a single large charge. The land Thumper (Geosource Inc. trade name) is a source of this type; its *charge* is created by dropping a heavy weight on the ground. In the continuous method, the source creates a relatively low-amplitude signal that may last 30 seconds or more. Data are recorded for the entire duration of the shot and then processed in a special fashion to make them look as if they had been shot by an impulsive source. The vibrator is a source of this type. Often several vibrators are used simultaneously by arranging them in an array pattern and synchronizing their signals in time. Thus, a vibrator source can be distributed in both time and space.

Seismic energy sources usually are compared to one another in terms of their strength or *energy level* because these parameters determine how well the sound from a given source will illuminate a given target. For land sources strength is stated as energy in megajoules, whereas for marine sources it is quoted in terms of the pressure in bars at a distance 1 m from the source. The table below shows the relative strengths for a selection of source types.

Table 3.1. Comparison of relative energy levels for various impulsive and vibratory sources.

		Energy (MJ)	Pressure (bar)
Impulse type			
Dynamite	1 kg	5.0	6.0
Thumper	3000 kg (3 tons) (1 drop @ 3 m)	0.08	land only
	30 drops	2.6	land only
Air gun	1639 cm^3, 138 bar	1.5	2.0
Gas gun	Single gun	1.3	1.5
Water gun	245.85 cm^3, 144.9 bar	marine only	1.0
Vibrator type			
Vibrator	56 HP, 14 s	0.6	land vibrator
	209 HP, 14 s	2.2	land vibrator
Mini-Sosie	1 impact	0.000135	land only
	1000 impacts	0.135	land only

Of the sources mentioned, only the gas gun is not in common use. A once popular source, the gas gun has characteristics inferior to those of the air gun, which replaced it in the early 1980s. The explosive sources are often restricted in use by environmental considerations. On land, explosives are not tolerated near population centers; offshore, explosives may damage reefs and kill fish. The air gun is by far the most popular marine energy source. The reasons for this are that its pulse is predictable and controllable, it has little impact on marine life, and it uses compressed air, a readily available substance.

The Society of Exploration Geophysicists (SEG) frequently publishes results of questionnaires sent to exploration companies. Such a survey conducted in 1990 showed that dynamite was the most commonly used land energy source. This result surprised many in the industry who thought vibroseis to be the most commonly used. However, the survey showed that the extremely high use of explosives in Africa outweighed the use of vibrators on a worldwide basis.

Table 3.2. Land Energy Sources in Use, 1989 (Goodfellow, 1990).

	%
Dynamite	51.18
Vibrator	46.82
Air gun	1.40
Gas gun	0.01
Thumper	0.17
Other	0.42

No equivalent statistics are available for marine sources, but by far the most used source is compressed air (sleeve and shuttle air guns).

The recording of acceptable-quality data is the most important consideration when choosing an energy source. Generally, an energy source is inadequate if the desired reflections are not observed on the recorded shot records. Usually this happens when the output of the source is too weak to penetrate to the target reflectors. Occasionally, a source is too strong. For example, a large air-gun array in shallow water with a hard bottom could create water-bottom multiples strong enough to exceed the dynamic range of the recording instruments (see Chapter 4). Once this occurs, any underlying signal is lost for good. Sometimes a source performing poorly can be improved by adjusting some of its operational parameters. For example, one can alter the characteristics of an air-gun array by changing the depth of the guns in the water, or one can tune up a vibrator array by selecting a more appropriate sweep function.

This chapter aims to give the geophysicist the information required to make sensible decisions concerning the selection, operation, and quality control of seismic sources. After a short section on source array design, land sources and then marine sources are described in detail. Some seldom-used sources are mentioned only briefly so that the most popular sources (i.e., air guns, vibrators, and dynamite) can be covered in depth.

3.1 Source Array Design

The theory of receiver array design applies equally well to source array design. Source arrays may be used to increase output energy level, cancel ground roll, and beam energy directionally. They provide the ability to improve the spectral characteristics of the outgoing energy if a concentration of energy in a particular bandwidth is desirable. This is one of the main reasons for the use of a source array.

Explosives, vibroseis, Dinoseis (Arco trade name), air guns, and Thumpers have been used in "patterns" or arrays. When using three or more sources in an array configuration, the resulting record is the received energy from a shot by that array. As the array moves along the line, individual records may be stored for later summing to produce an equivalent output as if the source array were longer or weighted.

On land, each source vehicle typically travels in a straight line with vehicles in either an areal formation or a linear formation. A staggered lineup may produce a weighted array. However, land source arrays tend to be more effective when relatively flat surface terrain conditions are encountered because, where difficult terrain is encountered, array formations often are disrupted during source movement around obstacles. This is one reason geophysicists sometimes prefer that arrays be simulated in the processing center rather than implemented in the field. Keeping source arrays simple also helps make field operations efficient.

When using both source and receiver arrays, each array response may be determined separately. The combined array response can be found by multiplication of the individual responses in the wavenumber domain. Figures 74, 75, and 76 show an example of a receiver array response, a source array response, and a combined array response for a horizontally traveling wave of different wave numbers. Note that the combined array response has better rejection of the horizontal wave than does either of the individual responses.

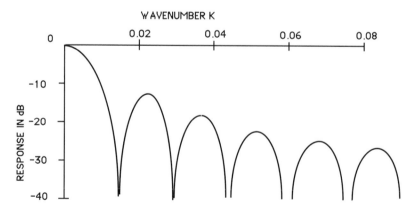

24 PHONES 2.7m APART

Fig. 74. Receiver response.

4 VIBRATORS 12m APART, 6 SWEEPS PER VP, 3m MOVE-UP

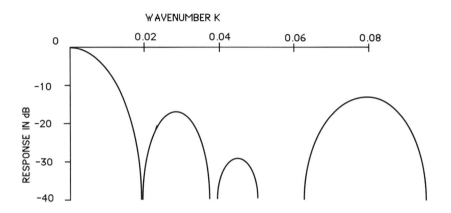

Fig. 75. Source response.

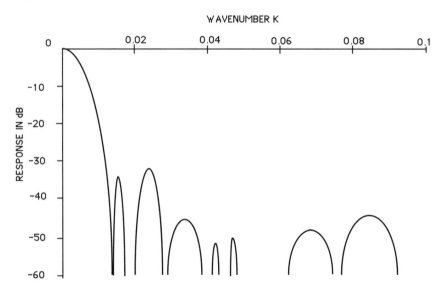

Fig. 76. Combined response.

3.2 Land energy sources

The choice of energy source is critical in land data acquisition because resolution and signal-to-noise quality often are limited by the source characteristics. A geophysicist should select a source based on the following five criteria:

1) **Penetration to the required depth**—Knowing what the exploration objectives are, the geophysicist should select a source that has adequate energy to illuminate the target horizons. Past experience can help here. If a five-element vibrator array produced good data in an area, one should hesitate to use, say, just a single vibrator in that area.

2) **Bandwidth for the required resolution**—If high-resolution reflections are required to delineate subtle geological features such as a stratigraphic trap, the source must transmit a broad range of frequencies, both high and low. For very shallow targets, a detonator may possess adequate energy and frequency bandwidth. For the deeper reflections, the longer travel path to a deep reflector requires the selection of a source that has enough energy at the higher frequencies to maintain a broad reflection bandwidth.

3) **Signal-to-noise characteristics**—Different areas have different noise problems. They may dictate the source selection. For example, areas having severe ground roll may require the use of source arrays to attack that problem. Also, the signal level created by a source can be area dependent. A vibrator running in an area of loose, dry sand may produce a weak signal because of poor coupling of the vibrator pad to the earth.

4) **Environment**—When working in populated areas, there are special safety requirements to which geophysicists must adhere. When vibrators are used in and around small towns, the survey lines may be diverted away from the town center to avoid causing traffic congestion or damage to buildings, flora, or fauna. Local authorities may not allow the use or transportation of dynamite in or near towns, and geophysicists must adapt to such regulations. In rural areas, any damage done to farm fields by vibrators or drilling rigs always must be repaired.

5) **Availability and cost**—The time of arrival of a crew can be extremely important. In such regions as the Arctic Sea and the North Sea, crews have to move into operation quickly to take advantage of favorable weather when it occurs. In land operations, there is no point in waiting months for special dynamite permits to be issued when vibrator crews may be available immediately, though at greater cost.

3.2.1 Land Explosives

Dynamite was the only source used until 1953, when the weight-dropping Thumper technique was introduced. Since that time, other land energy sources (vibroseis, air guns) have been introduced and have met with success in many areas of the world. Generally, dynamite produces more energy and a broader bandwidth than any other source. In-line or pattern charges (holes) have been useful in high-noise arid areas of Texas and North Africa. Large areal arrays consisting of holes that are 4.5 to 9 m (15 to 30 ft) deep and loaded with ammonium nitrate have been used with success. Dynamite, however, has some serious drawbacks that have led to the increasing popularity of alternative sources. Among those drawbacks are handling safety, high hole-drilling costs, logistical problems, hole blowout generating significant noise, and special crew requirements.

Most often, explosive land sources are detonated in drilled holes called *shot holes*. Figure 77 shows a drilling rig drilling the shot hole. The shot hole is loaded with an explosive charge, and the hole is back-filled with earth to stop explosive energy from escaping up the hole to cause a *blowout* and loss of energy when the charge is detonated. (Alternatively, in holes that collapse before the explosive can be loaded, the shot hole may be filled with foam and water to keep the shot hole walls firm for loading explosives.) The recording crew shooter comes along later, wires the explosive leads to an encoder known as a *shooting box* or a *blaster box*, and prepares the encoder to fire the charge by radio command from the instrument truck.

Compared to other land sources, explosives in shot holes have several advantages: easy portability to remote areas, high energy output, broad bandwidth, and simplified weathering corrections (since the shot is below the low-velocity layer).

Gelatine dynamite, nitrocarbonitrate (NCN), and ammonium nitrate are three types of land explosives commonly loaded down a shot hole.

Gelatine is a mixture of nitroglycerine, gelatin, and inert material. The mixture may vary in content depending on the charge strength required. The

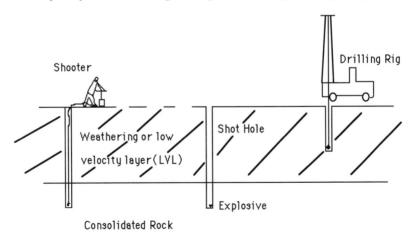

Fig. 77. The shot hole.

speed by which it explodes (i.e., the detonation velocity) is 6,500 m/s. The source creates a steep wavefront giving a broad-band, high-resolution pulse. Gelatine is not stable in storage for any length of time because the nitroglycerin separates.

NCN is more popular than Gelatine because it is less dangerous and less expensive. It is easier to transport; it is a class C explosive. A primer is used to set off NCN because of the NCN's low volatility. The primer may be 10 g of explosive with a velocity of 8000 m/s.

Ammonium nitrate granules are mixed with diesel fuel and poured into the shot hole and a primer is placed on top. This nitrate type has been used often in desert regions, where it is more economical to obtain supplies of such low-velocity explosives (it is a by-product of manure) than of high-velocity explosives.

The Gelatine and NCN types of seismic explosives generally are packed in plastic or cardboard tubes about 5 cm in diameter. The charge sizes vary from 0.5 to 5 kg per tube. The tubes often are designed to screw into each other for easy connection and lowering down the shot hole.

Detonators are used to initiate an explosion. A detonator is usually a small metal cylinder about 0.6 cm in diameter and 4 cm long. It contains a resistance wire embedded in powder. When a current passes through the wire, it heats up and ignites the powder, typically within 1 ms of the current application. A high-voltage *blaster decoder* (Figure 78) detonates the detonator or *cap*. The blaster charges a capacitor to a high voltage via batteries or a generator. The capacitor is discharged through the cap by radio command from the recording instrument encoder so that time reference zero is the instant the current is passed to the cap.

Detonators may be connected either in series or in parallel. When in series, all caps can be checked to ensure that they are electrically connected correctly. This is performed by placing a specially constructed detonator testing meter

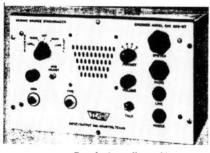
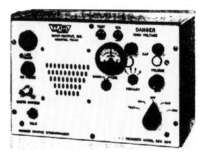

Encoder (recording unit)   Decoder (shot point unit)

Fig. 78. Blaster box (courtesy Input/Output Inc.).

across the ends of the series circuit. If a resistance less than infinity is observed (less than 2000 ohms in most cases), the detonators are wired correctly. A reading of infinite resistance means there is an open circuit and the charges will not fire (i.e., *misfire*). In a parallel circuit, an open circuit in the wiring or resistance element of any one of the detonators will not be observed readily because the total resistance being measured across a large number of caps will not change much if one of the parallel circuits is open. Hence, a series connection circuit often is preferred. However, there are reliability problems with a series connection; a break in the connection results in none of the caps firing. In the series case, either all caps fire or none fire, whereas in the parallel case, there is uncertainty about which caps may have fired.

Special techniques may be used to direct energy from the shot. *Shaped charges* (Figure 79) are used to break the rock during well-perforation operations (i.e., in order to release hydrocarbons from the formation) or during offshore trench-digging operations (prior to laying seabed pipelines). There has been some experimentation with shaped charges to examine their seismic beaming effect, but such experiments have shown little focusing of the energy any farther than a few meters away (Evans and Uren, 1985). Delay fuses and acoustic detonators are used in mine rock-blasting operations. Such detonators also have been tried in seismic exploration but are not used because their detonation velocities are low (around 3000 m/s) and the instant of detonation is unpredictable. However, mine quarry blasts have been used as a source in deep crustal refraction work where basin basement velocity information has been required.

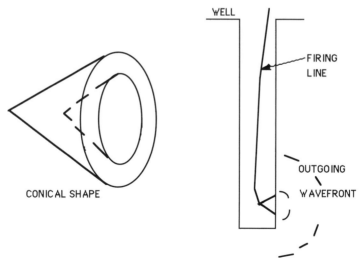

Fig. 79. Shaped charge.

In marsh areas, dynamite can be inserted into the soft ground using a tool called a *planting pole*. In other areas, shot holes have to be drilled. This may be accomplished by mud or air-circulation rotary drills, auger drills, or portable flush drills.

Dynamite does not necessarily require drilled holes. Primacord, an explosive in rope-like form, is sold under various trade names including primacord (Ensign Bickford Co. trade name) and Flexicord and Geoflex (Imperial Chemical Industries trade names). The cord varies in diameter from 0.5 cm to 1 cm. Primacord produces a broad-band, high-energy pulse and is packaged to allow easy ploughing to a meter underground. A detonator is taped to one end of the cord so the cord can be ignited (Figure 80). The cord can be supplied in weights of 10 to 40 g/m, depending on the firepower required. Generally, the seismic signal-to-noise ratio is influenced by the length of the detonated cord rather than the weight of it. It is apparent that the longer the cord length, the greater the signal-to-noise ratio, especially over areas containing high-velocity, near-surface layers, such as basalts (Evans, 1982). The problem with having long primacord explosives (in excess of 100 m) per shot is that, at a velocity of 7000 m/s, 100 m takes 14 ms to explode along its length. This causes the wavefront to expand slowly (see Figure 80), resulting in the wavefront energy output reducing from an energy spike to a broad pulse.

With a split-spread configuration, the detonator can be placed at the cord center so that the effects of the finite explosion time are symmetric on each side of the spread. If the dip is unknown, a center detonator with a split-spread configuration will ensure at least one spread is updip. The cord is buried with a plough (Figure 81) to a depth of at least 1 m in order to reduce air blast.

An alternative explosive energy source is the Betsy (Mapco Inc. trademark) for use on high-resolution surveys. A 21-mm shotgun pellet is fired into the ground vertically to produce a relatively weak, broad-band pulse. Small spatial patterns can be built from Betsy guns as shown in Figure 82.

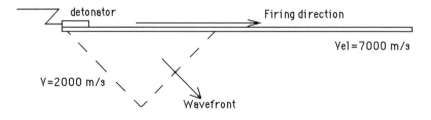

Fig. 80. Detonating cord.

Fig. 81. Primacord plow.

Fig. 82. Betsy three-gun carrier (courtesy Mapco Inc.).

Generally, a Betsy gun is effective for penetrating to a depth corresponding to about 0.5 s of traveltime.

3.2.2 Vibroseis

The original *vibroseis* technique was tested by Conoco in 1966 using a vehicle known as a *vibrator*. (The term "Vibroseis" is no longer a Conoco trademark.) The vibrator is a mechanical device mounted on a truck. Several trucks may be positioned along a line to make a source array. Vibrator trucks are normally as large as garbage trucks and weigh as much as 50 tons. They have the ability to move along public roads and sometimes have individual-wheel drive so they can be driven almost anywhere that can take their weight. The vibroseis technique is popular because the cost of operating vibrators is less than the cost of most alternative powerful energy sources. Only the Gas Exploder or "gas-gun" costs less, but its power is less than that generated by the vibrator.

The vibroseis concept is that if a signal containing a known set of frequencies is transmitted into the earth, the received signal at the surface, after extraction of the transmitted signal, will produce a signal that is the earth's reflection series. To transmit a signal containing a known set of frequencies into the ground, a steel plate is vibrated on the ground at known frequencies. This plate is the *base plate.* The vibrator truck is driven like a normal motor vehicle to the shotpoint, or *vibrator point* (VP) in this case, where the base plate is lowered to the ground. After the base plate completes its vibrating and the VP recording is completed, the base plate is lifted up and the vibrator truck driven to the next VP. This completes a single VP cycle, equivalent to an explosive shotpoint.

The base plate of a vibrator is driven by a continuous, variable-frequency, sinusoidal-like signal. At any particular time, the signal has an instantaneous frequency that lies within the seismic bandwidth. The driving signal is called a *sweep* because of the way in which the variable frequency sweeps through the seismic bandwidth. When the frequency range is swept from low to high frequencies, it is called an *upsweep*. A *downsweep* is a sweep from high to low frequencies.

To move the base plate at the desired frequencies, a driving force is required. The driving force is a heavy mass called the *reaction mass,* which is applied to the plate by hydraulic pistons. The movement of the hydraulic pistons is controlled by a *sweep generator.* The total hydraulic system, the reaction mass, and the base plate are mounted on the vibrator truck.

To ensure that the signal actually being transmitted into the ground is the desired sweep signal, a sensor like a geophone is mounted on the base plate. This sensor produces an electrical output signal that can be compared with the desired sweep signal. This electrical feedback loop keeps the transmitted signal the same as the sweep signal.

The method of recovering the reflection series from the signal received by the geophones is known as the *correlation process* (see Section 3.2.2.1). The sweep signal is often called the sweep, *reference*, or *pilot trace*. Once recording commences, a pilot trace is stored in the recording truck's electronic correlator unit. The received signal is stored in a memory unit, then correlated with the pilot trace. The output of this correlation process, known as the *correlated trace*, is saved on magnetic tape. A series of these traces, one from each geophone station, is a correlated field record, similar to a dynamite shot record.

In the field, the vibrator sweeps for a chosen period of time, known as the *sweep time*. The memory records the geophone output for the sweep time plus the time required to produce the desired record length. This additional time is called the *listen time*. An example of an upsweep and its recording cycle is shown in Figure 83.

3.2.2.1   Sweep Correlation

Figure 83 shows a reflected signal from three geological boundaries, each reflecting the complete sweep signal at a given time delay. Note that prior to correlation the signal at the detected surface does not readily indicate event arrival times. Two or more sweeps may be summed to build up the energy level and attenuate random noise. The summed signal is then crosscorrelated with the sweep signal. If a reflection event is present, a symmetrical wavelet is produced at a point in the correlated trace corresponding to the arrival time of that event.

Crosscorrelation of two series can be accomplished by reversing one of them in time and then convolving (see Chapter 2). For example, series (1, 3, 0,

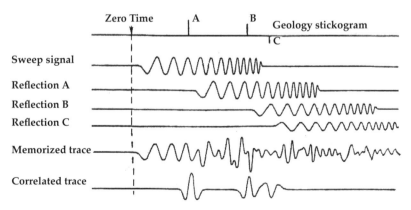

Fig. 83. Vibroseis sweep correlation (courtesy Conoco Inc.).

–2) crosscorrelated with (–1, 2, 2) produces an output (2, 8, 5, –7, –4, 2), which is the convolution of (1, 3, 0, –2) with (2, 2, –1). Thus, the correlated trace in Figure 83 is produced by crosscorrelation of the detected trace with the sweep signal. Note how the wavelets in the correlated trace occur at times corresponding to the presence of the three reflections in the geology trace. If the sweep signal is correlated with itself *(autocorrelation)*, a wavelet centered at time zero results.

In the field, the recording process is as follows:

1) Commence recording in memory and sweep for 8 s (for example), followed by another sweep (as many times as required).
2) Continue recording data in memory for as long as the final section is required (say 6 s) after sweeping ceases. The time from completion of sweeping to the end of recording is the listen time (Figure 84).
3) Vibrators move up to the next vibrator point as the correlation process takes place, and the final record is dumped from memory onto magnetic tape.

The vibrator sweep is governed by an electronic sweep generator. The sweep signal is transmitted to the hydraulic piston that moves the base plate. The output wavefront from the base plate is a function of the coupling between the plate and the elastic earth and, hence, a knowledge of the output wave form is of extreme interest for correlation purposes. Furthermore, vibrators are mechanical instruments subject to normal mechanical failure. Errors in sweep input to the earth must be monitored at all times so that they can be corrected immediately.

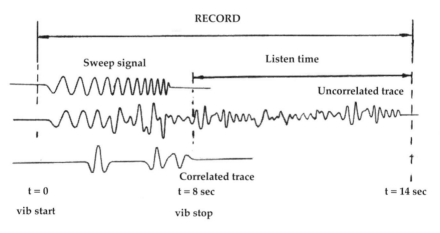

Fig. 84. Correlation process.

Because the output energy of a vibrator is spread over time compared with the spontaneous explosive effect of dynamite, vibrators can be used in populated areas (provided windows are well sealed since the vibrator energy may cause glass to vibrate and fracture).

Figure 85 shows a schematic of a vibrator system. The base plate is typically 20–25% of the reaction mass. The hydraulic piston causes the base plate to move relative to the mass. The vehicle's weight keeps the base plate coupled to the ground and, when the pistons move, the reaction mass motion is transmitted through the ground. The resulting ground motion depends on the driving frequency and the coupling with the earth. The base-plate mass attenuates continuous high-frequency motion. The flexible couplings prevent any high-amplitude low frequencies from being transmitted to the truck, which otherwise would shake it apart. Figure 86 shows a vibrator sweeping with its rear wheels off the ground to apply maximum weight to the base plate.

3.2.2.2   Sweep Control

One of the major problems with the vibrator or any other swept-frequency energy source is a lack of knowledge of the waveform transmitted into the earth. The pilot trace is what we would like to have transmitted because that is what is fed to the reaction mass. However, because of vagaries of the plate/earth coupling, the outgoing sweep frequency signature is never the same from one VP to the next.

Because the output beneath the vibrator base plate is not the same as the pilot trace driving the hydraulics, it is necessary to monitor the output frequencies and correct the driving electronics to try to maintain output as close

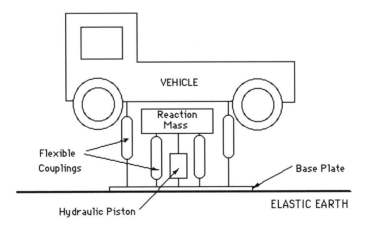

Fig. 85. Schematic of a typical vibrator sweeping in the field.

Fig. 86. Vibrator sweeping with its rear wheels off the ground.

to the pilot trace as possible. If the output is different from the pilot, the correlation process may fail and poor records could result. The correction technique involves tracking the changing phase of the sweep and pilot frequencies and making the output phase as close to the pilot phase as possible. This is known as *phase locking,* and the feedback comparator is called a *phase lock loop circuit.*

It is not possible to use a geophone to measure the base-plate motion because a geophone produces an electrical output proportional to velocity. Instead, since the base-plate motion is accelerating across a range of frequencies, we need a device that can give an output proportional to an acceleration movement. The instrument is an accelerometer and gives a raw output some $90°$ out of phase with the pilot signal. This is allowed for during correlation.

While sweeping across a range of frequencies, if the hydraulic drive makes the reaction mass move a greater distance than normal during selected frequencies, then there is an increase in the amount of ground force exerted by the vibrator base plate at those frequencies. This is known as changing the *drive level.* The more usual method of frequency control is to maintain a constant drive level and adjust the amount of time the sweep allocates to different frequencies.

An advantage gained by changing the drive level is that at desirable frequencies more energy may be imparted into the ground to enhance the recorded data at those chosen frequencies. However, this again is accompanied by changes in output sweep phase (Martin and Jack, 1990) and, consequently, it is hard to determine exactly what the true output sweep looks like, even with an accelerometer mounted on the base plate. One solution is to bury an accelerometer underneath the base plate and record its output instead of that of the base-plate accelerometer. However, this is not feasible in a production sense, so we have to accept the base-plate accelerometer as the best indication of output motion.

3.2.2.3   Sweep Design

The sweep frequency range is designed to obtain the highest reflection signal-to-noise ratio and data quality/character. The vibroseis sweep is programmed to transmit a broad range of frequencies. Shallow reflectors require a broad bandwidth containing high frequencies. Deeper reflectors do not, in general, return higher frequencies because of the absorption properties of the earth. One solution to higher-frequency absorption is to concentrate sweep frequencies in the higher end of the spectrum. Initial parameter testing is necessary to observe the results. Often, such tests show that the effort is not rewarded with the result desired and the effort expended is not considered worthwhile. Hence, the inclusion of less high-frequency energy is often the case in sweeps designed for deeper targets, compared with vibrator sweeps designed for shallow horizon reflections.

An upsweep with a constant sweep rate (in Hz/s) is known as a *linear sweep* in time. The frequency spectrum of data recorded with a linear sweep normally has most of its energy in the lower frequencies, generally as a result of poor base-plate coupling and attenuation of the high frequencies. As an alternative choice, a nonlinear sweep may make the received frequency content more equal across the spectrum. A nonlinear sweep may be logarithmic (spending more time at lower or upper frequencies). Other sweep types (time-squared and amplitude-tapered) are possible, but generally the linear or logarithmic sweeps are accepted as the most useful. Linear sweeps provide an equal time spent at all frequencies, useful when ground roll is not an extreme problem. Logarithmic sweeps allow more time to be spent at the higher end of the frequency spectrum, enabling the enhancement of high-frequency reflected energy while requiring that less time be spent generating low-frequency ground roll.

The sweep type is selected for the area of operation. The propagating wavelet after correlation is zero-phase. The greatest resolution wavelet results from the broadest sweep range. Figure 87 shows how the correlated wavelet depends on sweep parameters. To the seismic interpreter, high-amplitude side lobes (the negative troughs on either side of the central positive peak) are unwanted because they make horizons difficult to interpret. Ideally, a large-amplitude, high-frequency peak is preferred because it offers improved resolution and thin bed definition. A short-duration, high-amplitude spike with low-amplitude side lobes is the seismic interpreter's preference. Having as broad a bandwidth as possible is preferred to varying the center frequency. In

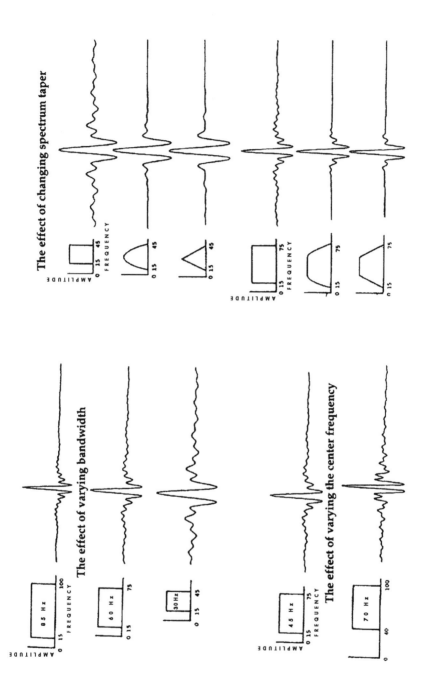

Fig. 87. Bandwidth, frequency, and taper.

Figure 87, a 60 Hz bandwidth sweep is shown with center frequencies of 45 and 70 Hz. The wavelet has a narrower time duration at the 70 Hz center frequency, but the lower 45 Hz center frequency wavelet does not introduce deleterious side lobes while the higher-center frequency does. Thus, interpretation is easier if the bandwidth is wide enough to avoid the introduction of excessive side lobes. The center frequency should be chosen at a value where the most useful frequencies for interpretation are received. Figure 87 also shows how sweep tapering affects the received wavelet.

The design of operational sweep-frequency parameters can depend on many factors, both geophysical and operational. For example, higher frequency sweeping may not increase the bandwidth of the received signal because the higher frequencies may be absorbed by the earth anyway. Above 100 Hz, the vibrator often is difficult to synchronize, so the range often is limited in practice to below 100 Hz. Vibrators built recently can have operating ranges as broad as 5–250 Hz. Although lower frequencies are not absorbed as readily as high frequencies, low-frequency sweep generation below 20 Hz also can cause operational problems, such as an increase in vibrator hydraulic line failures. Consequently, when designing sweep-frequency parameters, one should understand the limitations of frequency transmission as a result of absorption as well as the operational impediments involved in sweeping the desired range. Once again, compromises are often reached to ensure acceptable signal-to-noise ratio results while retaining efficient operations.

Ideally, the vibrator force should be as large as possible, the sweep range wide, and the sweep length long in order to impart the maximum amount of broad-band energy into the earth. Thus, signal-to-noise ratio improves as force, sweep length, and bandwidth increase. As was discussed in Chapter 1, the signal-to-noise improvement in the stacked section is approximately a square-root relationship to the increase in fold. In vibroseis, a rule of thumb is that the signal-to-noise improvement is approximately in square-root proportion to the increase in force, sweep length, and bandwidth. That is,

$$\frac{\text{Signal}}{\text{Noise}} \text{ Improvement } \alpha \ \sqrt{FLW}, \tag{30}$$

where F = force, L = sweep length, and W = sweep bandwidth.

The noise referred to here is the noise generated by the force. Signal tends to increase at a different rate than noise during increasing force so that the signal-to-noise ratio does not remain constant by changing the force alone. If the predominant noise is environmental background noise, then equation (30) is not a reasonable approximation. Doubling sweep length theoretically has the same effect as doubling bandwidth on signal-to-noise improvement. Thus, since bandwidth may be limited, longer sweeps or a greater number of recordings may compensate. Two 8-s sweeps are equivalent to one 16-s sweep in signal-to-noise improvement, but in practice, the delay between two sweeps makes one sweep preferable.

3.2.2.4   Side-Lobe Noise

As shown to in Figure 87, a square frequency spectrum produces unwanted side lobes in the autocorrelation wavelet. These can be reduced by tapering the spectrum. If the sweep amplitude spectrum is tapered off at frequencies at both ends of a sweep, the result is to reduce side lobes, but this also reduces energy. A compromise is to taper the sweep from 0.0–0.3 s or 0.0–0.5 s on each end to reduce side-lobe energy while maintaining the majority of the desirable energy and frequency content.

Cunningham (1979) proposed side-lobe reduction by use of a pseudo-random coded signal, impressed on the transient vibrator sweep, that would allow full power on the vibrator sweep without tapering. However, this approach does not appear to have been adopted by the industry.

Alternatively, attenuation of side lobes can be attained by varying the sweep-frequency range between successive sweeps (known as Varisweep). If a vibrator sweeps across six different frequency ranges and bandwidths, the correlated signal will have different side-lobe wavelengths. These then could be summed to produce a correlated signal containing little side-lobe energy, as shown in Figure 88.

Therefore, in practice, instead of all vibrators having the same sweep frequency range, vibrators may be programmed to sweep different spectra at different vibrator points (VPs), as shown in Figure 89. This sometimes is referred to as *frequency modeling* the data.

Here, three vibrators are shown schematically as they move from right to left with vibrator 1 in the lead. The programmed frequencies of each vibrator at each VP are shown.

Here, the vibroseis array is not only weighted in position but in frequency as well. It is possible with some vibrators to achieve an equivalent frequency weighting by changing the ground force as a function of frequency, rather than by changing the sweep frequency range. Summation of records builds the peak and attenuates the lobes as shown in Figure 88. Crosscorrelation is performed before summing.

The Varisweep sweep can be up or down, linear or exponential (the latter causing frequency spectrum shift and, thus, side-lobe attenuation). If the low-frequency spectra of the sweeps is concentrated at the outer edges of the array, there is the potential for horizontally generated ground roll cancellation as the sweeps are summed after crosscorrelation. Hence, it may be preferable

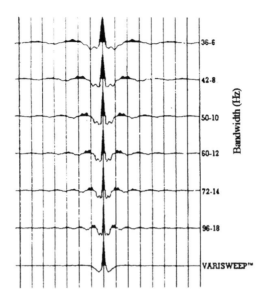

Fig. 88. Varisweep (courtesy GeoSystems).

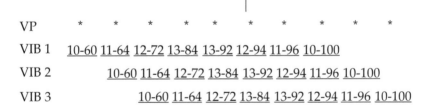

VP	*	*	*	*	*	*	*	*	*	*
VIB 1	10-60	11-64	12-72	13-84	13-92	12-94	11-96	10-100		
VIB 2		10-60	11-64	12-72	13-84	13-92	12-94	11-96	10-100	
VIB 3			10-60	11-64	12-72	13-84	13-92	12-94	11-96	10-100

Fig. 89. Vibrator frequency modeling.

to program the swept-frequency range so that the lower frequencies are swept at the ends of the vibrator array and the higher frequencies are spread across the array's center. In practice, a reduction in low-frequency ground roll can result from sweeping the lower frequencies at the ends of the array. This can cause the frequency spectrum to become squarer than it is when large-amplitude ground roll has been generated by low-frequency sweeping at the center of the array. An example of this is Figure 90, which shows the spectrum of an actual field record after vibrator frequency modeling was used to produce a relatively flat peak of power across a frequency range of 14–38 Hz.

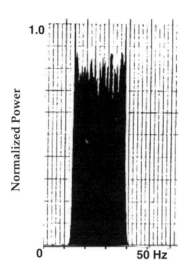

Fig. 90. Power spectrum of recorded data after
frequency modeling (courtesy GeoSystems).

3.2.2.5   Vibrator Harmonic Ghosts

If the vibrator base plate is poorly coupled to the earth (i.e., when sweeping high frequencies over inelastic ground or vibrating over hard rock), it temporarily might lose contact for parts of the sweep. The loss of contact can cause base-plate warping at particular sweep frequencies, which can cause secondary waves to be transmitted. These are called *harmonics* of the original sweep frequencies. As a result, the outgoing vibrator sweep transmitted through the earth is not the same as the pilot sweep and, hence, the crosscorrelation process does not work correctly. Such false waves are known as *harmonic ghosts,* which appear on recorded data.

There have been two methods proposed for the elimination of harmonics: (1) using upsweeps of the desired sweep time and frequency, and (2) phase shifting the sweep frequency during recording.

Using Upsweeps

With a downsweep, the sweep (i.e., the pilot trace) crosscorrelates at a positive time shift, producing a harmonic ghost at positive record time. Figure 91 is a good example of a downsweep and the resulting harmonic ghosts (and side lobes). The two-way traveltime of a harmonic ghost is given by

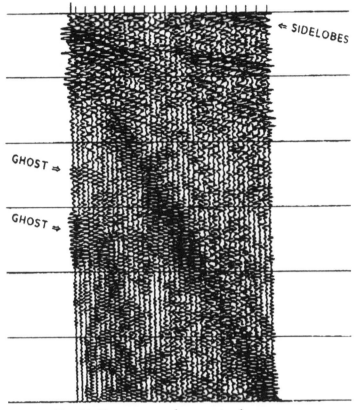

Fig. 91. Downsweep harmonic ghosts.

$$\Gamma = \frac{Lf_1}{f_2 - f_1},\qquad(31)$$

where L = sweep length, f_1 = lowest sweep frequency, and f_2 = highest sweep frequency.

For a 12-s downsweep from 60–15 Hz, an initial ghost begins at 4 s, and further ghosts may appear at later times. Ghosts have the same moveout as other arrivals, but the time of their appearance on the record can be changed by changing the sweep length and/or by reducing the sweep bandwidth.

With an upsweep, the crosscorrelation process causes the harmonic ghost to appear at a negative correlation shift, hence at a negative two-way travel-time. Therefore, by recording upsweeps, these harmonic ghosts correlate before zero time and so do not appear on the record.

Phase Shifting the Sweep Frequency

The phase-shift approach (Rietsch, 1981) aims to reduce the amplitude of the first harmonic of the sweep frequency because that harmonic is usually larger than the others. The method works by recording a number of consecutive sweeps with different phases at each sweep location and then summing their results to produce a harmonic-free radiated signal.

For example, in Figure 92a, an input downsweep trace is shown with its first harmonic, which is twice the input sweep frequency. The radiated signal from the bottom of the base plate is shown beneath them. It is the sum of the sweep and its first harmonic.

In Figure 92b, the same radiated sweep signal (but the full length) is shown as trace 1, where the harmonics cause extreme deterioration in the radiated signal. Traces 2, 3, and 4 are the same trace shifted 90°, 180°, and 270°, respectively. Trace 5 is the difference between 1 and 3, showing a reduction in harmonic noise. Trace 6 is the difference between 2 and 4. Trace 7 is trace 5 shifted by 90° and summed with trace 6 to give greater attenuation of the harmonic noise. This process, which requires electronic modifications to the sweep generator, can attenuate the harmonics from a vibrator.

Of the two approaches to harmonic elimination, the upsweep method is the most popular. Even when vibrator sweep generators have been modified

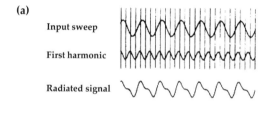

Fig. 92. (a) Input sweep signal and its first harmonic. (b) Results of shifting and summing individual sweeps.

to incorporate sweep phase shifting, the vibrator crew tends to record upsweeps just to ensure there are no ghosts generated on records.

Summary of Vibrator Problems and Possible Solutions

Here is a summary of vibrator problems and possible solutions:

1) A long sweep may cause additional vibration to equipment, thereby causing faster deterioration of equipment. Long sweeps can reduce side lobes and improve resolution and signal-to-noise ratio. However, this is at the expense of reduced operational speed, which may not be financially acceptable. A decision must be made on whether to accept reduced signal-to-noise ratio or to improve this ratio, but at a price.

2) A broad bandwidth is required for the best seismic resolution possible. A frequency range should be selected to maintain ground roll at a minimum level.

3) Downsweeps are easier to control than upsweeps at low frequency. This is because the reaction mass displacement is inversely proportional to the drive frequency and it is easier to start with low displacement and increase accordingly. However, downsweeps can generate harmonic ghosts, which are only eliminated by special modification to sweep generator electronics, or using upsweeps instead. Sweep length should be adjusted to remove harmonic ghosts.

4) Often any improvement in signal-to-noise ratio is proportional to the increase in the number of sweeps per VP. The signal-to-noise ratio can be improved by increasing the effort expended, a function of force, sweep length, and bandwidth.

3.2.2.7   Starting a Vibroseis Crew

This section outlines the basic techniques for determining the field parameters that would optimize a vibroseis exploration crew's recording efficiency and data quality in an area where no previous recording has been performed. The parameters that need to be determined and the techniques used to determine them are not necessarily unique to the vibroseis system of exploration. They all have a direct or analogous counterpart in any source method used in CMP reflection work. The parameters that need to be determined are:

1) Coverage of receiver group and source
2) Minimum and maximum offsets (split spread or off-end)
3) Receiver group interval
4) Number of geophones per group
5) Number of sweeps per source point and number of vibrators to be used

6) Spectrum of the sweep
7) Length of the sweep
8) Source point interval
9) Amount of CMP fold

Coverage of receiver group and source—The configuration of the source pattern and geophone coverage, irrespective of the number of sweeps given per source point or the number of geophones in a receiver array, is an initial requirement. Often, the distances over which the source and geophone groups are spread are made equal to help attenuate the horizontally traveling Raleigh waves, air waves, and other surface waves. Generally, in areas where in-line coverage is used, the length of the source and geophone group is set equal to one wavelength of the highest velocity surface wave that is to be attenuated. If the frequency of the surface wave is not known, a good rule of thumb for determining this wavelength is to divide the velocity of the wave by 10. The best way to obtain the velocity is to record a noise test or a noise spread.

Minimum and maximum offsets—The minimum and maximum offsets of the geophone spread need to be determined, as does whether the setup should be an off-end or split spread (see Chapter 2).

Receiver group interval—The group interval must be determined. This is the distance between the center of one group and the center of the next adjacent group. This parameter is related to the choice of offsets, the number of recording channels available, and possible cable limitations (see Chapter 2). It is also a function of the steepest dip expected in the area, the maximum frequency expected, and the target reflector velocity.

Number of geophones per group—The number of geophones per group usually has an upper limit because of the number physically available to the crew (see Chapter 2). This number always should be sufficient to handle almost every case in which a crew finds itself operating.

Number of sweeps per source point and number of vibrators—The number of vibrators, the number of geophones, and the number of channels available on the recording instruments affect the spatial sampling. The number of sweeps per source point and how the data are processed later affect the signal-to-noise ratio of the data. The number of vibrators available is usually fixed, but the only limitation on the number of sweeps is the time required to record each source point or profile.

Spectrum of the sweep—The spectrum of the sweep must be determined so that most of the vibrator time is spent generating useful frequencies rather than frequencies that are not reflected or lost in noise. In the case of Varisweep, after vibrating a series of different frequency sweeps, the data may be summed. A computer capable of performing a Fourier transform produces the resultant spectra, and the aim then is to enhance the amplitudes of reflection frequencies. The resulting combinations of sweeps should show that target reflections are enhanced. A bandwidth of at least two octaves is preferred in field data, though that may be impractical in deeper reflected data.

Length of sweep—The preferred sweep length is usually a function of the harmonic ghosting that may be encountered. This is a consideration when dealing with downsweeps. The computation for the ghost time is given by equation (31). For example, according to that equation, a 54–6 Hz downsweep lasting 16 s has a ghost time of 2 s. If some target reflections are expected to occur later than the 2-s ghost time plus the first arrival time, the sweep parameters must be changed to get a longer ghost time. In general, the two parameters that affect the ghost time the most are the length of the sweep and the low-end frequency of the sweep. In a situation in which the possible ghost is too shallow, the low-frequency end of the sweep should be raised or the length of the sweep increased. Because the low-frequency end of a sweep may be selected for other reasons (see the spectrum of the sweep paragraph above), in most cases we may be reluctant to raise it. Therefore, the best parameter to change to get a later ghost time is the length of the sweep. This is one reason the sweep lengths, particularly in areas where the low frequencies are of interest, become fairly long, in the neighborhood of 13–18 s. Consider eight sweeps of 20-s duration, each with a 5-s listen time. This would take 8 x (20 + 5) = 200 s per vibrator point. However, 16 sweeps of 10-s duration with the same listen time would take 16 x (10 + 5) = 240 s. Hence, longer sweeps are clearly more efficient; this is a further reason for preferring longer sweeps.

Source point interval—The source point (vibrator point or VP) interval is normally a function of the receiver station interval. When using more than one vibrator in line, the VP is considered to be at the center of the vibrator array. Noise spread tests determine the distance vibrators must be apart to minimize ground roll; the shortest vibrator array length is bumper-to-bumper. The source interval is dependent on the signal-to-noise ratio and the amount of coverage required to enhance signal-to-noise ratios. Also, see the discussions on field coverage in Chapter 2.

Amount of CMP fold—The CMP fold is determined before the field work begins. If the survey is in an area that has been recorded successfully with vibroseis previously, it is useful to decimate the fold (by reprocessing the existing data) to observe if a reduction in fold still can give an acceptable seismic section. If a reduction can be tolerated, then the acceptable level of fold will determine the source interval. If the survey is in an area where little vibroseis has been recorded previously, then one practice often adopted is to start recording with a CMP fold of half the number of available recording instrument receiver channels. If data quality is poor, then all channels are used. However, if data appear acceptable after the stack has displayed good-quality reflections, then further fold reduction may be made.

3.2.3 Other Land Energy Sources

3.2.3.1   Thumper

The Thumper or "weight dropper" is a popular source in desert areas (Figure 93). It was introduced by McCallum in 1953 and used widely in Saudi Arabia and central Australia. It generates a high level of horizontally traveling surface waves and, relative to other impulsive sources (such as explosives), is of low energy level. Typically, the weight is a two-ton steel block, dropped from a height of 3 m (10 ft). The records from multiple drops are summed. Summing multiple drops is successful provided the weight does not bounce, causing secondary waves to occur and be summed. Weights frequently bounce, depending on the terrain upon which they are dropped; this is a shortcoming of the technique. The chains hanging down around the Thumper weight are a safety measure meant to prevent field hands from walking underneath the weight during operations.

In early Thumper surveys, large geophone groups or *patches* were used with two Thumpers, one at short offsets and the other one at long offsets (Figure 94). The geophone patches were designed to reduce ground roll. The patches were often geophone arrays arranged in a star-like pattern, with the takeout at the center. The Thumpers worked at different offsets because the limited number of stations that could be connected to the recording instruments meant that coverage had to be improved by more weight drops,

Fig. 93. Thumper weight-drop vehicle (courtesy Geosource Inc.).

increasing the number of source points. By weight dropping on each side of a patch, additional traces could be added to produce longer offset split-spread records, with the spread becoming symmetric when all the individual traces are summed to produce one record.

By comparison with yesteryear, today's operations have many more channels available, so that when two Thumpers are used, one drops while the other moves up. A record is produced every 10 s during good operational periods when all equipment functions correctly. Figure 95 indicates how one or more Thumpers are arranged on line to record as a source array. Moving along diagonally side by side, they each drop at the dashed points shown, the dashes representing individual drop points per Thumper spread. The individual records are summed to produce Drop 1, Drop 2, etc., so that in this figure, Drop 3 centered at station 3 is also duplicated by segments of Drops 2 and 4. This acquisition scheme usually produces good reflection coherence and some ground roll noise attenuation.

Thumpers rarely are triggered to drop at the same time because the distance from the weight-release point to ground level can vary with the terrain over which the Thumpers work. If they were to drop in unison, the weights

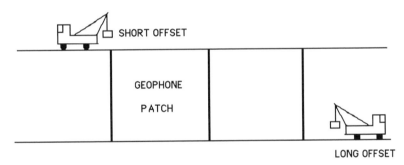

Fig. 94. Offset recording with two Thumpers (plan view).

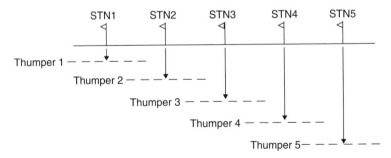

Fig. 95. Thumper array recording.

hitting the ground would have to arrive only milliseconds apart to avoid loss of high-frequency content in the total outgoing signature.

3.2.3.2   Dinoseis

Sinclair introduced the Dinoseis technique in 1962. The method uses the explosion of a propane-oxygen mix in an enclosed chamber to blast a movable bottom plate against the earth. Prior to detonation, the heavy chamber sits on the ground. Upon explosion, the chamber flies upward and is caught by a mechanism to avoid it dropping back to the ground to produce a secondary impact. Unlike with the Thumper, three or four Dinoseis units can be fired simultaneously (Figure 96). The advantage of the Dinoseis is found in resupply. Bottles of propane and oxygen used by the Dinoseis are generally readily available worldwide. However, the Dinoseis signal was weak relative to that of vibroseis, which has taken most of the work that was available for Dinoseis operations.

3.2.3.3   Air Gun

One version of the land air gun consists of a steel bell filled with water and sealed at the bottom by a diaphragm that rests on a base plate. An air gun is mounted within the bell and filled with air to 138 bar (2000 psi). When the air gun is fired, the air bursts into the water, causing expansion of the diaphragm which transmits the pressure wave to the base plate. In reaction, the bell is driven upward clear of the base plate and is caught 50.8 cm (20 inches) above the base plate. It is then slowly lowered back into position for the next shot. As with vibroseis, a number of units (three or more) may be fired simultaneously. An output from base-plate-mounted sensors is transmitted back to

Fig. 96. Propane–oxygen gas gun (courtesy Teledyne).

the recording truck to give a firing time signal (Figure 97). For country use, the air gun may be mounted on the back of a tractor.

Alternatively, a specialized air gun may be made to operate within an auger drill, which may be drilled into soft ground (see Figure 98). A shot with such an air gun has a minimum air blast. Auger air guns have been used in marsh type areas where other forms of energy source have had operational problems. Care must be taken to ensure there is no invasion of mud into the air chambers.

3.2.3.4 Mini-Sosie

The Mini-Sosie (Figure 99) uses a multiple-impact, pedestrian-controlled rammer and is equivalent to a mini-vibroseis (requiring correlation and summing). Similar rammers can be seen on building sites for ramming the earth in preparation for load-bearing construction.

The Mini-Sosie ground force is little more than the sum of the rammer weights as they vibrate up and down impacting on the ground. A sensor mounted on the base plate (or "shoe") records each impact, which is controlled by the speed of a two-stroke engine piston. The sensor output is stored in sign-bit format (see Chapter 4). At the end of recording, a crosscorrelation with the sign-bit-recorded data produces the seismogram.

The Mini-Sosie suffers from occasional poor coupling like the vibrator, but two or three rammers may be linked for use. This provides better signal output and a limited potential for random noise attenuation, since the Mini-Sosie is essentially a random signal generator. The rammer is walked from one station to the next. Two rammers are often used side by side to increase reflection signal levels.

Mini-Sosie is used as a high-resolution, shallow seismic energy source in mineral, groundwater, coal-mining exploration surveys and engineering studies.

3.2.3.5 Hydrapulse

A Hydrapulse (Figure 100) is a rammer that uses hydraulics to maintain its ground force (rather than the simple weight dependency of the Mini-Sosie). Its output energy is higher than that of the Mini-Sosie, but it has lower repetition rates. Source arrays of two or three tractor-mounted machines may be used where terrain access is difficult for other sources (such as in jungles or forests).

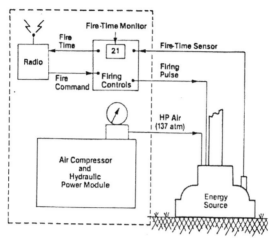

Air gun source flow diagram

Fig. 97. Air gun (courtesy Bolg Associates).

Fig. 98. Dual-auger air guns.

Fig. 99. Mini-Sosie (courtesy CGG).

Fig. 100. Hydrapulse (courtesy C.M.I. Corporation).

3.3 Marine Energy Sources

As with the land energy source, desirable characteristics for a marine energy source are: (1) maximum-output signal-to-noise ratio, (2) high-output energy, (3) high resolution, (4) minimum disturbance of the environment, (5) low capital and maintenance cost, (6) convenience of resupply, and (7) reliability. In water, the available energy sources for use can be classified as explosive, implosive, and vibrator types.

When an explosive source (such as dynamite) detonates in water, the explosion causes a gas bubble to form. The outgoing exploding wavefront provides the pulse of energy needed for the seismic survey reflections to occur, but the bubble formed by the remaining gases becomes a significant problem as a generator of secondary wavefronts, which are considered coherent noise. The bubble expands until its gas pressure is less than the hydrostatic pressure of the water head around it. The bubble then begins to collapse, compressed by the hydrostatic pressure. Bubble compression continues until the pressure reaches a level almost as great as the pressure created during the initial explosion. The bubble again expands and collapses, producing another outgoing wavefront. This can continue for several cycles. The oscillating bubble slowly rises in the water because of the differential waterhead pressure between its top and bottom. As Figure 101 shows, the rate at which the bubble rises depends on its radius. The result of this oscillating bubble is that each recorded reflection event is represented by a whole sequence of pulses—a primary followed by many *bubble pulses*. As the bubble pulses make interpretation of a seismic section difficult, they are considered a

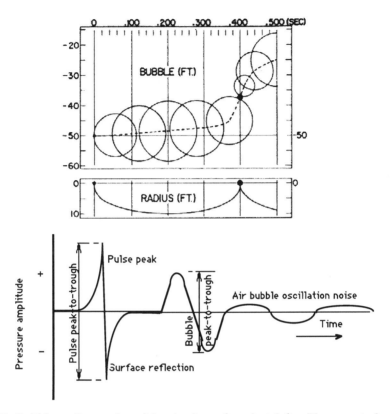

Fig. 101. Bubble radius and position in time after shot (after Kramer et al., 1968), and a simplified schematic of its pressure signature.

The shape of the pressure signal created by a marine source is called its *signature*. Figure 101 shows a generic signature for an explosive marine source. The initial positive peak is the explosive pulse. This is followed by a trough that is the reflection of the initial pulse from the sea surface. Following this is the bubble-generated noise train. The useful signal is measured by evaluating the peak-to-trough amplitude, whereas the harmful noise train maximum peak-to-trough amplitude is evaluated as the noise or "bubble" level. The performance of an energy source is rated by the source signature pulse-to-bubble ratio and the pressure output 1 m away from the source upon detonation. The pulse-to-bubble ratio is an indication of how close the signature is to that of an ideal pulse, and the pressure output is evaluated in bar-meters. Because the pulse-to-bubble ratio is an initial indicator of purity of the source signal, it is therefore desirable to attain as high a ratio as possible.

The source signature can be affected by depth of detonation and other factors in the source design. Source signature testing and evaluation is discussed in detail later in this chapter.

The bubble problem affects all types of marine explosive energy sources. Approaches to resolving this problem include the use of implosive energy sources in which the outgoing wavefront is generated by the collapse of a cavity in the water and the marine vibrator that sweeps rather than generates a pulse of energy. Vibrator sources are seldom used in marine surveying, whereas impulsive energy sources such as air guns are used most commonly.

3.3.1 Air Guns

Air guns are the most common type of marine impulsive energy source. There are two main types of air guns—*shuttle guns* and *sleeve guns*.

3.3.1.1   The Shuttle Gun

The principle of the air gun is that a high-energy pulse occurs by the sudden release of a volume of high-pressure air. Consider Figure 102, which shows a section through a shuttle gun. The gun has two air chambers with a centrally mounted piston or *shuttle*. The shuttle moves between both chambers; a hole through its center allows the passage of air from the upper chamber to the lower chamber. High-pressure air (138–345 bar, 2,000–5,000 psi) is

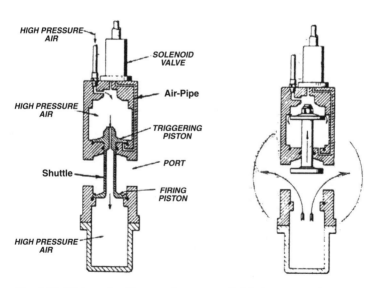

Fig. 102. Firing shuttle gun (courtesy Bold Associates).

supplied to the upper chamber; this flows through the shuttle to fill the lower chamber with air.

A solenoid is positioned at the top of the upper chamber to control air to the *air pipe*, a thin pipe that leads from the upper high-pressure chamber through the gun casing to the bottom of the shuttle's upper flange. The solenoid is connected by an electrical line to the seismic vessel. The air gun is supplied by air (via air hoses or "lines") from air compressors aboard the seismic vessel. To release the high-pressure air into the water through holes in the air-gun assembly, an electrical impulse to the solenoid allows a short pulse of air to escape along the air pipe.

When the air pulse arrives underneath the shuttle flange, an imbalance of air pressure to either side of the flange dislodges the shuttle to move it up and slightly farther into the upper chamber. As this occurs, air is released from the bottom firing piston end of the shuttle, which has been blocking the flow of air from the lower chamber. The air then bursts out of the lower chamber through port holes in the assembly casing (Figure 102). After most of the air is exhausted, the high-pressure air feeding to the top chamber forces the shuttle back down to seat above the lower chamber, thereby filling both chambers. At this point the gun is ready for the next shot.

The air-gun energy is a function of the pressure and the volume of the air stored in the air chambers. Shuttle-gun volumes typically range from 164–4920 cm^3 (10–300 in^3); air-pressure levels commonly used in the industry range from 34.5–345 bar (500–5000 psi).

3.3.1.2   The Sleeve Gun

An alternative gun type is the sleeve gun, in which there are fewer parts (and, hence, fewer equivalent maintenance requirements). The sleeve gun allows air to escape in the form of an air annulus rather than via the four ports of the shuttle gun (Figure 103). Both measurements and theory indicate that the amplitude of the initial pressure pulse produced by an air gun is proportional to the square root of the gun's port area. Since the outlet annulus area of a sleeve gun is significantly larger than the port area used by a shuttle gun of similar volume, a sleeve gun produces a more powerful source signature. Nevertheless, the signatures of a sleeve gun and a shuttle gun (of the same volume) have the same spectral shape. Hence, the two guns have equivalent frequency bandwidths.

Individual guns may be chained to each other so that they no longer function as individual source points but as arrays or strings towed behind the seismic ship. Guns also may be arranged in rigid groups or subarrays, either for towing astern of the ship or from arms extending outward from either side of the ship.

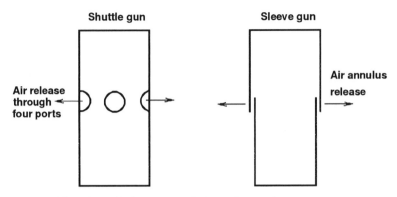

Fig. 103. Shuttle and sleeve gun firing schematics.

3.3.2 Sparker

The sparker generates a pressure pulse by the discharge of a high electric current between electrodes in the water. The electrical energy is stored on board in capacitor banks. The discharge generates heat in the water, thereby vaporizing it and creating a steam bubble like a dynamite explosion.

3.3.3 Flexotir

The Flexotir (CCG trade name) source creates a pressure pulse by detonating a 56.7-g (2-ounce) explosive charge in a cage, which is a cast-iron perforated spherical shell with a diameter of 0.6 m (2 ft). The charge is pumped down a hose to the cage and fired with a delayed detonator. After detonation, the cage dissipates the bubble, which helps to reduce gas bubble oscillations (Figure 104). The cage cannot be used in shallow water because the Flexotir cage may become entangled in seabed objects and break away from the hose.

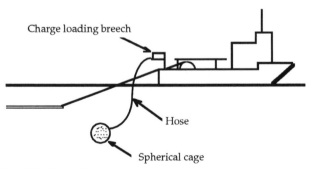

Fig. 104. Flexotir.

3.3.4 Maxipulse

Maxipulse (Western Atlas Inc. trade name) utilizes a special cartridge (the size of a soft drink can) containing 200 g of NCN and a delay detonator (cap) that protrudes from the top of the can. Water is used to flush the can down a hose trailing behind the seismic ship. One end of the hose is mounted on the stern of the ship, and the other end of the hose trails at about 10 m below the sea surface. At the deep end of the hose is a rotating wheel or "shoe." When the cap end of the can strikes the wheel, the cap detonates after a 1-s delay. This delay allows the can to move away from the hose so that the explosion does not damage the hose. The explosive has a large bubble pulse, so the signature for each shot must be recorded for later debubble processing. This is done by positioning a pressure detector (hydrophone) on the hose end. The first bubble pulse is often higher in amplitude than the initial pulse from the explosion.

3.3.5 Detonating Cord

A detonating cord (otherwise known as primacord or Geoflex as discussed previously in this chapter) is thin plastic tubing filled with explosive, which is supplied in coiled drums. It often is used where the water depth is too shallow to allow standard seismic vessel operations. A detonator is taped to one end of the tube as in land operations. The cord is placed in the water from small boats, wired to a normal blaster unit, and radio-fired remotely. A detonating cord has a minimal bubble because the rope-like cord floats just beneath the sea surface. Detonating cords have been used in environmentally sensitive areas with good results seismically and environmentally.

3.3.6 Gas Gun

The gas gun source, otherwise known as Aquapulse (Western Geophysical trade name), generates a pressure pulse by detonation of a propane–oxygen mixture (Figure 105). A flexible rubber "boot" is filled with the mixture. When ignited by a car spark plug, the mixture explodes and causes the boot to expand rapidly. The ballooning of the boot causes the initial pulse, followed by secondary oscillations as the boot returns to its original shape. These oscillations are reduced compared to those resulting from uncontained explosions. The exhaust gases are vented to the water surface. The more the boot is worn, the worse the secondary oscillations become. The output energy is considered poor compared with that from other sources, but this source produces acceptable results in some areas, and propane–oxygen bottles are relatively cheap. One benefit has been the reduction in multiples—observed in some areas of Indonesia and northwest Australia—compared to those that result from other

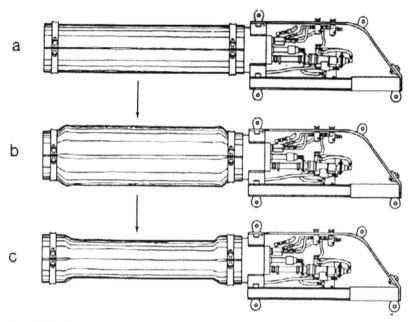

Fig. 105. Oxygen–propane gas gun system.

sources. This reduction may be a result of the spectral content of the gas gun signature, which is not as broad-band as that of an air gun. Usually four guns are fixed in tandem.

3.3.7 Water Gun

The water gun uses a shuttle driven by compressed air to move water through a chamber (Figure 106). When the shuttle is at rest, it is held in position by compressed air. A firing signal opens a solenoid valve, causing a rapid increase in air pressure on one side of the shuttle, which immediately responds by moving down the water chamber. This causes a high-velocity water plug to eject through four ports. A momentary void forms in the water behind the jets. Implosion of the void then occurs, causing a single energy pulse output. The water guns are operated at about 138 bar (2000 psi), similar to air guns. Their energy levels are comparable to those of air guns. Unlike the air gun, a water gun has no oscillating air bubble. However, it does have several disadvantages. As with the Vaporchoc steam gun described below, the signature has an undesirable precursor peak. Furthermore, the implosion timing depends on gun depth. If the water guns are fired at variable depth, then array-gun, firing-time synchronization becomes a problem.

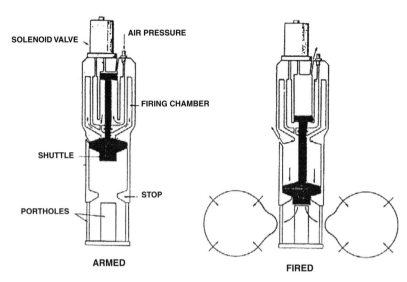

Fig. 106. Water gun.

3.3.8 Steam Gun

The steam gun (otherwise known as the Vaporchoc, Figure 107) was developed by CGG and is one of the impulsive energy source types used in marine exploration. A volume of superheated steam is injected under pressure into the water. This generates a pressure pulse like that produced by an air gun when it fires. As the steam bubble expands, its pressure decreases until it is below the hydrostatic pressure. Implosion then occurs according to Rayleigh's law,

$$t = 0.91 R_o \sqrt{\frac{W}{P_o}}, \tag{32}$$

where t = implosion time (s), R_o = maximum bubble radius (cm), W = water density (g/cm^3), and P_o = hydrostatic pressure (bars).

The implosion causes a second shock wave, with the pressure reaching thousands of bars. This wavefront is the main outgoing pulse for seismic reflections. The source signature consists of two unequal amplitude pulse peaks, 1 and A, separated by bubble period T (Figure 108).

Ideally, the most desirable marine source signature has a high spike-shaped pulse and no coherent noise caused by bubble oscillations. The Vapor-

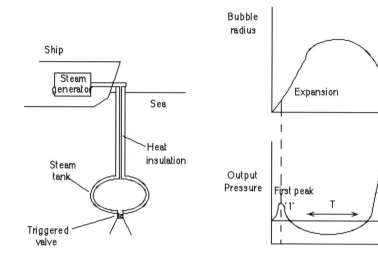

Fig. 107. Vaporchoc schematic   Fig. 108. Vaporchoc firing schematic.
(courtesy CGG).

pulse and no bubble. A signature minima occurs at frequencies $1/2T, 3/2T, 5/2T$, etc. The greatest energy transfer occurs at frequencies $1/T, 2/T, 3/T$, etc.

During survey operations, a steam boiler and super heater produce three tons of steam per hour, supplying 197 000 cm³ (12 000 in³) at 69 bar (1000 psi) every eight seconds. Steam is fed down a supply line to a tank or "gun" in the water. A steam-triggered valve has outlet jets controlling the steam injection into the water. Up to eight individual outlet jets may be employed, and two guns are used (one being spare). The steam temperature is 320–400 °C. The implosion peak amplitude is independent of temperature but proportional to steam pressure supplied. Thus, if greater output peak amplitude is required, the steam pressure is increased, which upon supply to the tank increases output jet volume and, hence, steam bubble size. When the system is operated at maximum pressure, considerable maintenance is required. That is one of the system's disadvantages.

The bubble period T is related to steam pressure as follows:

$$T = Kp^{-5/6}, \tag{33}$$

where K = system constant and p = steam pressure.

The signature amplitude can be adjusted some 12 dB by changing pressure. As with the water gun, the implosion period T depends on gun depth in the water. In practice, pressure peak A varies in inverse proportion to the jet exit length. The initial peak can be reduced by decreasing the steam jet length, but this then reduces output energy.

3.3.9 Flexichoc

The Flexichoc source is a chamber consisting of two steel plates joining four expanding sides. The plates initially are separated by compressed air and locked into a fixed position. The air is evacuated and, shortly afterward, a port is opened rapidly, causing an inrush of water and implosion with no subsequent bubble. Compressed air is then injected into the chamber and the plates locked into position, followed by evacuation before the next shot. Flexichoc was developed by CGG and is rarely used today.

3.3.10 Vibroseis

Two versions of marine vibroseis initially were tested during the late 1980s by Geco-Prakla and Western Geophysical. The vibrators were mounted on a sled towed behind the recording vessel. Vibrators also could be towed in arrays at 5–7 m depth, and some comparisons with air-gun arrays showed very similar reflection–signal strength and frequency content. The vibrators ran with a continuous sweep (typically 10 s), with no listen time between sweeps, but were correlating as the vessel sailed along the seismic line. While comparisons were reasonable when compared with air guns, marine vibrators had to be designed so as to avoid cavitation, which could cause radiated signal deterioration and potential correlation failure.

The marine vibrator may have an application in offset vertical seismic profiling. The Geco parent company, Schlumberger, investigated this approach.

3.4 Marine Air-Gun Arrays

Individual air guns have a signature that is a function of their air volume, pressure, and operational depth (Figure 109). During the conception of the air-gun array in the early 1960s, only two or three guns with a total volume of 9830–14 700 cm^3 (600–900 in^3) at 82.8 bar (1200 psi) were used. Arrays currently use 15 to 100 guns with a total capacity ranging from 23 800–147 000 cm^3 (1450–9000 in^3). Typical air pressure is 138 bar (2000 psi). Experimental arrays have had a total air volume as high as 197 000cm^3 (12 000 in^3).

Since a fixed air pressure and gun depth are operationally desirable, air volume and the total number of guns employed are the variables of an air-gun array. The use of many air guns in the form of an array provides an increase in outgoing energy and also provides a means of reducing the ampli-

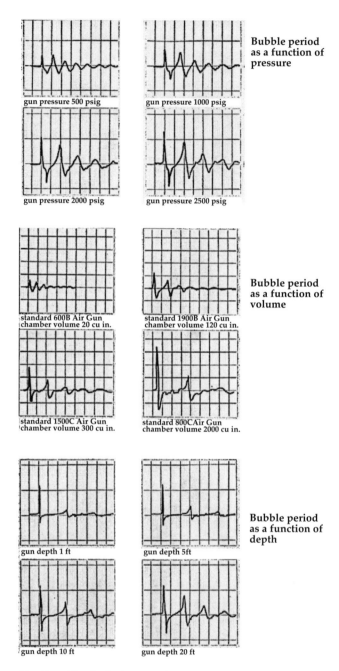

Bubble period as a function of pressure

gun pressure 500 psig

gun pressure 1000 psig

gun pressure 2000 psig

gun pressure 2500 psig

Bubble period as a function of volume

standard 600B Air Gun chamber volume 20 cu in.

standard 1900B Air Gun chamber volume 120 cu in.

standard 1500C Air Gun chamber volume 300 cu in.

standard 800C Air Gun chamber volume 2000 cu in.

Bubble period as a function of depth

gun depth 1 ft

gun depth 5ft

gun depth 10 ft

gun depth 20 ft

Fig. 109. Typical air-gun signatures.

tude of secondary bubble pulses. When air guns of different volumes are chosen so as to optimize the array's primary-to-bubble ratio, the array is said to be *tuned*. When two or more air guns in close proximity are fired as a *cluster* (replacing individual large guns), the bubbles may coalesce into one, increasing the primary-to-bubble ratio when compared to an individual gun. For example, a two-gun cluster can have a larger initial pulse amplitude and a lower bubble amplitude than a single gun of the same volume as the cluster. Better tuning is possible using so called cluster guns. Individual groups of guns may be arranged in subarrays so that, on firing, all gun signatures are added as shown in Figure 110.

When an air gun receives a firing pulse, its mechanical components begin to move against each other at different speeds (friction causes a time delay at the start of movement, known as *stiction friction* in mechanical engineering). Consequently, the actual firing time of individual guns tends to vary. Repeatability of timing is critical since all guns must fire in synchronization to produce the signature of a correctly tuned array. A fairly wide industry standard adopted in the past has been to ensure that no guns fire more than 2 ms before

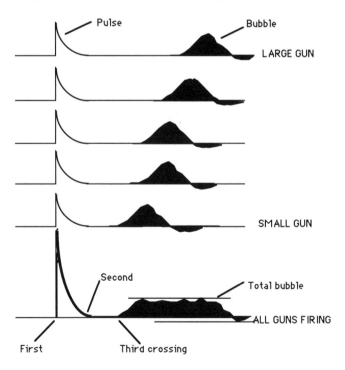

Fig. 110. Five guns summed to improve pulse-to-bubble ratio.

or 2 ms after the array's average firing time. Guns failing this criterion are considered defective. This can happen because of normal wear and tear of electrical and mechanical parts. All guns are monitored independently by sensors on each gun body that detect shuttle movement. These transmit a pulse to the monitoring equipment on the vessel. Hence, any firing-time delays required to synchronize the guns may be executed aboard the vessel without retrieving the guns. Such movement sensors also detect any malfunction or misfires by the guns.

With the aid of computer programs, arrays can be designed for ideal summation of pulses using subarrays of guns towed in various configurations. Guns may be towed in two strings (just like geophone arrays) behind the seismic vessel (Figure 111) or in individual subarrays from booms spread out across the stern of the vessel. An alternative to towing from a boom is the use of *paravanes*, which are submerged tubes with onboard steering mechanisms. The paravane has the advantage of remote steering control from the ship so that the positions of individual subarrays with respect to each other and the

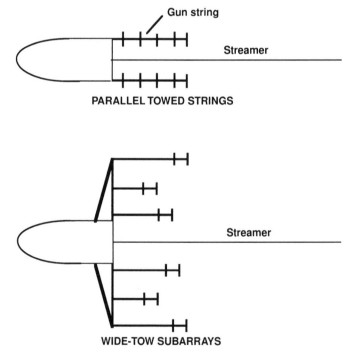

Fig. 111. Gun-array tow configurations.

vessel may be changed as required; the use of booms alone restricts this capability. Some vessels may use a mixture of booms and paravanes.

Subarrays of guns spread broadly across the line to be recorded are termed *wide-tow arrays*. When all of the guns are fired at the same time, the cross-line array response cancels some *side-scattered energy* (which is reflected or refracted from surfaces out of the plane of the line being surveyed) or *back-scattered energy* (that energy being reflected/refracted back toward the streamer). Such wide-tow arrays should be used only where side- and back-scattered energy exists. They should not be used in areas where the geology has dipping horizons across the line direction, reflections of which may be canceled by the wide-tow array response. Such back-scattered noise may be stacked out during conventional CMP stacking (Lynn and Larner, 1989). The industry has tried a variety of long, wide, and different geometry arrays. Clearly, the shorter the array width, the less scattered the noise attenuation, but the less attenuation of steeply dipping reflections.

One additional advantage of using wide-tow subarrays is the ability to record more than one seismic line at any time. It is possible to fire individual subarrays alternately to record two or more separate midpoint lines per ship traverse. These configurations are discussed in Chapter 7.

3.4.1 Far-Field Testing

The performance of an energy source such as an air-gun array is of major importance in data acquisition. This is monitored by executing a *far-field signature test*, which provides:

1) Peak-to-peak (p-p) amplitude of the primary energy pulse measured in bar-meters (bar-m), equivalent to the outgoing pressure wavefront 1 m from the source
2) Pulse-to-bubble ratio defined as the ratio of primary-pulse peak-to-peak amplitude to bubble peak-peak amplitude after the signature's third zero crossing (Figure 110)
3) The optimum air-gun array design, including how an array performs when individual guns fail to fire

The performance of an air-gun array may be measured in the field by positioning a single hydrophone directly below the array. This has been done by towing a depressor from a ship's anchor cable and trailing a hydrophone from the depressor as shown in Figure 112. This is known as a far-field signature test. The opposite of this type of test is a *near-field signature test*, in which each gun's signature is observed close to each gun (hydrophone strapped to a gun-tow chain). To be in the far field, the hydrophone has to be far enough from the source array so that all source-to-hydrophone raypaths differ by no more than about one recording sample interval. Consequently, the hydrophone should be located at least 300 m below the array. Furthermore, the seabed must be at least 200 m below the far-field phone to prevent contamination of the measurement by sea-floor reflections. In practice, this may be hard to achieve because the water depth may not approach 500 m for many hundreds of nautical miles offshore (meaning the seismic ship must travel for days to arrive at an acceptable depth). Consequently, compromises must be made. Far-field tests have been performed in 200 m of water. In such cases, the hydrophone may be as close as 50 m from the nearest gun in the array. Results from such tests are extremely questionable.

A problem associated with towing a depressor beneath the gun array is that if the results are to be representative of the far-field signature, the calibrated hydrophone ideally should be maintained directly beneath the array. This is a difficult exercise to perform operationally because the ship must move continually forward at normal recording speed, and any crosscurrents that cause the hydrophone to drift away must be countered by the ship steering a course at an appropriate speed into the currents. This may require the ship to travel at very slow speeds (not at the normal recording speed of about 5 knots).

Signatures also have been measured using sonobuoy telemetry systems, which allow the vessel to move at normal recording speed without the problems of towing a depressor and hydrophone, as shown in Figure 113.

The advantage of telemetry transmission is that sonobuoy hydrophones are less noisy compared to towed hydrophones. Also, when the hydrophone is towed, the vessel speed must be maintained at a low level to keep the hydrophone at depth. This can cause gun depth variation and, thus, changes in the array's signature. However, the sonobuoy method has the disadvan-

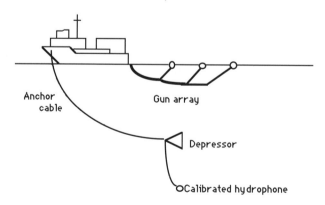

Fig. 112. Far-field signature ("pulse") test.

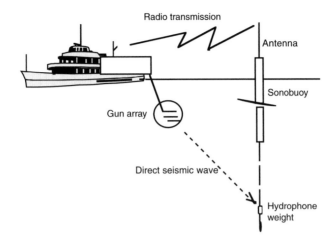

Fig. 113. Sonobuoy far-field recording technique.

tage that the source-to-hydrophone geometry is not fixed and, hence, every recorded shot may appear different because of array directionality.

The hydrophone output is recorded onboard the seismic vessel in the form of a voltage waveform that represents the source signature. The hydrophone sensitivity must be known (typical sensitivity is 0.1–2 V per bar) accurately prior to the test. (In the United States, hydrophone calibration can be verified at the U.S. Navy research laboratory in Orlando, Florida.) The one-way distance of the source from the receiver is obtained from the one-way traveltime (difference in time between firing of the guns and arrival reception) by multiplying this value by the local velocity of sound through water.

The computation of the signature wavefield amplitude is then

$$P = \frac{VD}{S}, \qquad (34)$$

where P = outgoing pressure in bar-meters (pressure at a distance of 1 m from source), V = peak-to-peak volts measured, D = one-way distance, and S = hydrophone sensitivity.

3.4.2 Shot Timing

Multielement sources should fire at precisely the same time. To compensate for inherent mechanical delays, firing times are adjusted electrically. Normally, each gun has a movement sensor built within or adjacent to the gun's chamber. The sensor produces an electrical signal at the instant of firing. The timing for each gun is then observed by an onboard computer, and the electrical firing time of each gun is adjusted so that all guns fire within a predetermined interval. On older gun controllers, the firing time for each gun was adjusted manually on the run-in to the start of each line so that all guns would fire at the same time when the line recording began. With firing controlled by computers, the use of a staggered firing command has been investigated with the aim of simulating beaming effects of the energy source. This method has not yielded outstanding results.

Individual gun-firing times may be displayed either on paper or on video. The advantage of hard copy is that it can be kept for record and quality control (QC) purposes.

3.4.3 Relative Energy Source Levels

It is difficult to compare relative energy source levels and pulse-to-bubble ratios because they are dependent to a great extent on both the source depth and the recording technique. However, while only being approximate in value, these figures provide some insight as a comparison:

Table 3.3. Relative energy source levels.

	Bar meters	Primary-to-bubble ratio
Aqua pulse-4 Gas guns	12	4:1
Maxipulse	10	0.9:1
20 air guns (27 044 cm^3 138 bar)	35	8:1
40 air guns (65 560 cm^3 138 bar)	75	>10:1
Vaporchoc	6	3:1

Figure 114 shows the Raleigh-Willis diagram relating bubble period to potential energy for a number of sources, expressed in logarithmic scales. Small energy sources (such as a detonator denoted as "det.") have low energy but relatively high frequency when compared to larger sparker units. The larger energy sources usually have the greater bubble periods. Sources with longer bubble periods tend to have lower bubble amplitudes. Note that individual air guns fall midway between sparkers and explosives. No recent work has been performed to update this diagram using water guns or different air-gun arrays.

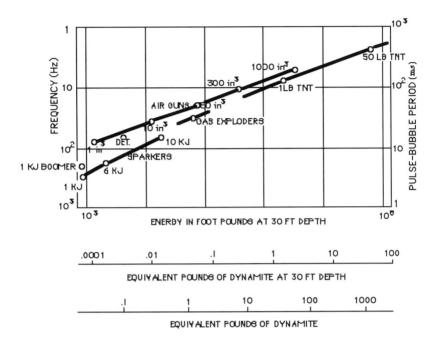

Fig. 114. The Raleigh-Willis diagram relating pulse-bubble period to potential energy.

3.5   Source and Receiver Depth (Ghost Effect)

On land, the burial depth of a dynamite charge can affect the exploding wavefront's amplitude and shape. Tests have been conducted with charges loaded in clay, sand, water-filled holes and cemented holes over the years (e.g., McCready, 1940). The frequency spectrum may increase with depth but can be distorted by the surface ghost. Shallow charges often have poor amplitude and frequency content because of detonation within a porous weathering layer. Ideally, the charge should be placed beneath the weathering for improved statics corrections and superior signal-to-noise ratio, plus less surface noise.

The charge depth governs a phenomenon called *ghost interference*. As shown in Figure 115, a ghost is created by the downward reflection of the primary pressure pulse from the surface, the weathering layer, or both. A ghost has a polarity opposite to that of the primary.

If the ghost arrival time corresponds with a true reflection, the shot depth must be adjusted immediately. This tends to be more of a problem with land, where hole depth may be greater than 30 m (100 ft), than marine, where air-

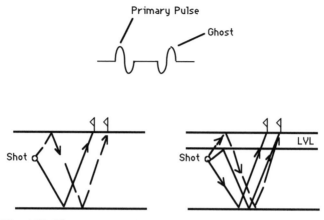

Fig. 115. Ghost generation.

the ghost interfering with a reflected event but the ghost actually causing notching of the frequency spectrum.

In Figure 116, the solid line represents a direct downgoing wave and the dashed line represents the wave reflected from the surface. The physics of sinusoidal wave propagation states that when two waves have the same wavelength, destructive interference (cancellation) occurs when they arrive exactly 180° out of phase. Because the reflection coefficient at the surface is negative, the downgoing reflected wave experiences a 180° phase shift relative to the direct wave. However, the reflected wave experiences a further phase shift because of the additional distance, $2d$, that it travels relative to the direct wave. If destructive interference is to occur, that distance must be an integral number of wavelengths. That is, destructive interference occurs when $2d = n\lambda$, where $n = 0,1,2,....$ Since $\lambda = V/f$, the notch frequencies where destructive interference is experienced are given by

$$f = \frac{nV}{2d}, \ n = 0, 1, 2, \dots . \tag{35}$$

For example, if $V = 1500$ m/s and $d = 6$m, then source ghost-notch frequencies are at 0, 125, 250,... Hz. Figure 117 shows field examples of the signature of an air-gun array as the array changes depth. PTP refers to the peak-to-peak strength, and the PBR refers to the primary-to-bubble ratio.

Figures 118 and 119 display normalized versions of the normalized amplitude spectra for source and receiver arrays as a function of source and receiver depths. Figure 118 maintains the source array constant at 5 m and

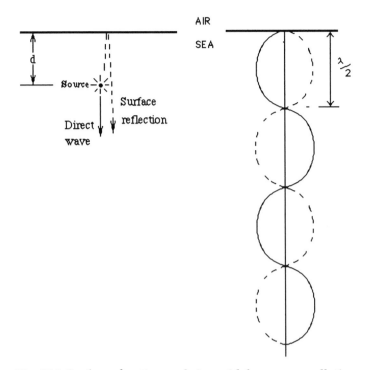

Fig. 116. Surface ghosting and sinusoidal wave cancellation.

changes the streamer depth from 5 to 15 m, while Figure 119 repeats the exercise but with the source array at 10 m. These examples show how streamer depth can affect the location of notches in the spectra and how important it is to maintain a constant source and receiver depth.

The ghost notches are not of infinite depth because of noise in the recorded signatures and their finite length. In particular, the signature truncation produces a finite DC component (at 0 Hz). This has the effect of making the ghost notch that actually occurred at $f = 0$ appear instead at about $f = 3$ Hz.

Ideally, a streamer should be towed at a depth designed to minimize the impact of the receiver ghosts on the spectrum of the seismic data. At depths of less than about 6 m, the ghost notch at $f = 0$ begins to seriously attenuate the low end of the seismic spectrum. At depths of 15 m or more, the first nonzero ghost notch affects the higher end of the spectrum. A 10-m streamer depth is a reasonable compromise that has become something of a de facto standard for streamer surveys.

For land work, the ghost becomes a real problem if good-quality recording requires the shot to be placed beneath a thick weathering layer. Ghost notch-

ing can then occur as a result of reflections from beneath the weathering and
from the surface.

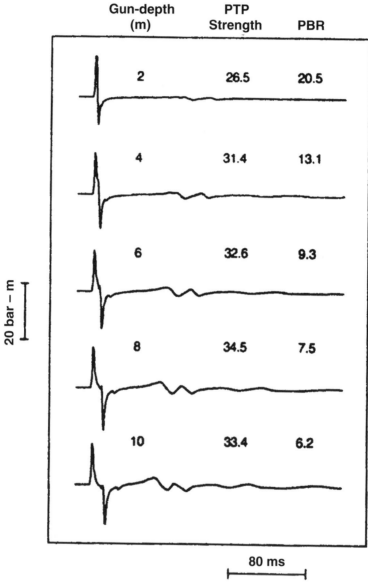

Fig. 117. Air-gun array signatures for a varying depth of source
(after Dragoset, 1990).

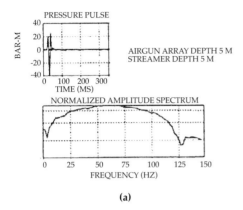

(a)

Fig. 118. Array
response, source at
5 m, receiver depth
variable (courtesy
Geco-Prakla).

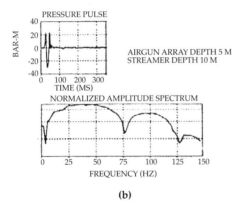

(b)

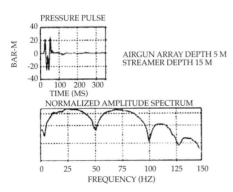

(c)

Signature and normalized amplitude spectrum
(a) with array at 5 m depth and streamer at 5 m,
(b) with array at 5 m and streamer at 10 m,
and (c) with array at 5 m and streamer at 15 m

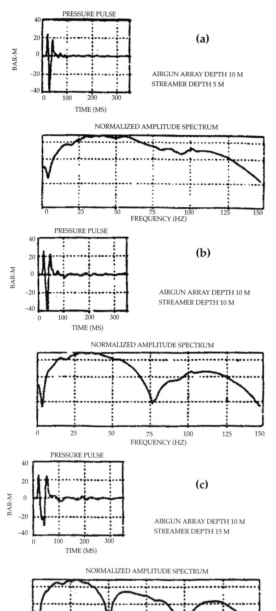

Fig. 119. Array response, source at 10 m, receiver depth variable (courtesy Geco-Prakla).

Signalization and normalized amplitude spectrum (a) with array at 10 m depth and streamer at 5 m depth, (b) with array at 10 m depth, and (c) with array at 10 m depth and streamer at 15 m depth.

3.6   Determining Optimum Air-Gun Specifications

Although the seismic industry typically uses signature time-domain characteristics to characterize air-gun arrays, this is an inadequate approach to determining array performance. Because time-domain characteristics can be modified by filtering, time-domain specifications (i.e., peak-to-peak and pulse-to-bubble ratios) are only valid when accompanied by a description of how the air-gun signature was filtered. An array signature recorded with a 2-ms sample rate (125 Hz antialias) cannot be compared in the time domain with another recorded with 1-ms sample rate (250–375 Hz antialias). The shorter sample interval recording will exhibit greater peak values because it has twice the bandwidth. Instead, amplitude spectra are preferred for comparison purposes: Any false impressions caused by sampling will be evident on the amplitude spectra. Amplitude spectra should be referenced to a fixed minimum pressure. Energy-source researchers use micropascals (μPa), equivalent to 10^{-11} bars.

The total number of guns provides the greatest influence on output energy, not the pressure or the total array capacity. Two guns having a total volume of 1639 cm^3 (100 in^3) will give a greater output than a single 3280-cm^3 (200-in^3) gun. This is because the signal output strength is proportional to the cube root of each gun's volume. For example, if the two guns are each 819 cm^3 (50 in^3), their net strength is proportional to $2(50)^{1/3}$ while the strength of the single gun is proportional to $(100)^{1/3}$. The ratio of the strength of the two guns to that of the single gun is $2(50/100)^{1/3}$, which is approximately 1.6. An excellent treatise is offered on the subject by Dragoset (1990). For those who need to check gun specifications before the start of a survey, this is mandatory reading.

Exercise 3.1

Select the energy-source parameters for a vibrator survey with the following objectives in mind:
- Target depth—3000 m
- Velocity at target—3000 m/s
- Optimum resolution desired, using a sweep length of 16 seconds

Consider Figure 120 while solving the following problems:

1) Using the signal-and-noise spectra, select the preferred sweep frequency range to minimize the generation of ground roll but maximize the reflected signal. In your selection, compromises will have to be made.

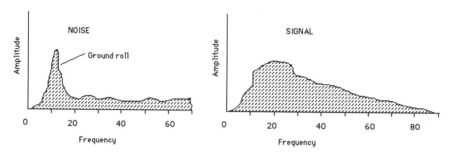

Fig. 120. Signal-and-noise characteristics from a previous dynamite survey.

2) If a downsweep were used, what time would the correlation noise occur on the record? Would this interfere with the target reflection?
3) Field tests have shown that a single vibrator record has an S/N of 2 to 1 at the target depth when a sweep of 8-s length and 20 to 60 Hz is used. What combination of sweep length and sweeps per VP would give a 4-to-1 improvement in signal-to-noise ratio?
4) If 10 s are needed for the moveup between sweep points, how long will it take one vibrator to record one VP? Assume a 6-s listen time, a basic sweep of 8 s, and an S/N improvement of 4.

Exercise 3.2

The direct energy wave and its ghost provide a two-point filter operator and hence discriminate against certain frequencies (see Figure 121). Sketch the frequency response for the two source depths and decide which depth gives the broadest frequency range with the minimum number of notches in the seismic bandwidth.

Exercise 3.3

In a real situation, a marine survey is being conducted with a 2-ms sample rate and the hi-cut filter is 125 Hz. Sea state has deteriorated and, in order to maintain a noise-free cable, the vessel must reduce speed to 4 knots and dive the cable to 13.7 m (45 ft). Unfortunately, the front end of the air-gun arrays dips to 13.7 m (45 ft) and the back end is maintained at 10.6 m (35 ft) by buoys. Sketch the source frequency response. Assuming the cable is now noise free, would this be adequate justification to continue recording? If not, why not? What would be the effect on the frequency response if the source were maintained at 7.6 m (25 ft) but the receiver (streamer) were put at 15.2 m (50 ft)?

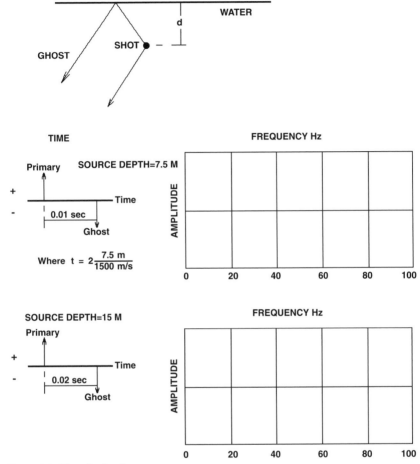

Fig. 121. Sketch the frequency response.

Exercise 3.4

The following data were provided by a crew as part of the results of a pulse test:

- Velocity of sound through the water = 1500 m/s
- Amount of hydrophone cable reeled out = 250 m
- Gun depth = 6 m
- First arrival traveltime = 10 ms

What can be said of the location of the hydrophone?

Chapter 4

Seismic Instrumentation

4.1 Introduction

So far, the complexities of the seismic source and receiver have been discussed. Once a seismic signal is transmitted and received, it must be recorded. The different types of signals discussed in this chapter are defined as follows:

1) *Source signal*—The pressure field created by the seismic source.
2) *Reflectivity signal*—The earth's reflection sequence convolved with the source wavelet.
3) *Seismic signal*—Everything received as a result of the source firing. The seismic signal includes the reflectivity signal as well as ground roll, refractions, diffractions, sideswipe, channel waves, etc.
4) *Received signal*—The electrical output of the receiver group. This is the seismic signal plus all environmental noise.
5) *Recorded signal*—The data, that is the instrument-filtered signal plus any additive instrument noise, which go on tape.

The information contained in a signal can be characterized by three quantities: signal-to-noise ratio, bandwidth, and duration. Signal-to-noise ratio can have different meanings depending on the circumstances. For example, diffractions from out of the reflection plane are noise on 2-D data but are part of the signal in 3-D surveys. In seismic exploration, the recorded signal bandwidth is usually 0–250 Hz or lower. Often, data are processed in a narrower band, say 5–80 Hz, even though they may be recorded in a broader band. The duration of recorded signals depends on the nature of the source and target depth. Impulsive sources, such as land dynamite or marine air guns, create a source signal with a duration of a few tenths or hundredths of a millisecond. On the other hand, the duration of an uncorrelated vibrator source signal may be 10 s or longer.

A reflection is a physical event caused by a change in the acoustic impedance of the earth. In a recorded signal, that event is represented by a wavelet

159

that has two components—the earth filter and the acquisition wavelet. The wavelet can be described in the time domain or, alternatively, in the frequency domain. The Fourier transform can be used to move from one representation to the other. If a wavelet has a short extent in time and appears like a spike, it is likely to be composed of a broad band of frequencies, each separate frequency having its own phase value. The amplitude and phase of a wavelet contains all the spectral information of a wavelet. These spectra are called the *frequency-domain representation* of the wavelet, whereas the wavelet in time is considered to be in the *time domain*. For a brief explanation of the frequency domain, the reader may refer to the Appendix.

4.2   Basic Concepts

When seismic recording first began in the 1920s the recording systems consisted of heavy, metal-cased geophones connected by wire cables to a recording truck. The geophones were analog sensors that transmitted analog signals to analog instruments inside the truck. Each incoming analog signal was connected to its own analog amplifier, after which the signal was recorded on a rotating photographic drum (similar to the equipment still used today at many seismic observatories). Drums were replaced by analog magnetic tape recorders (like the reel-to-reel tape recorders sometimes used by radio stations) during the late 1950s, but these often failed to operate well. In the early 1960s they were replaced by digital tape recorders, each of which had an analog-to-digital converter at the input to the tape drive. The individual analog amplifiers also were unreliable, and by the late 1960s, they were being replaced in recording devices by a single multiplexed analog amplifier. This single digital amplifier allowed the incoming analog signal from all geophones to be recorded digitally.

The amplifier required a multiplexer switch to synchronize the switching of geophone lines to the amplifier at the correct time. In the late 1970s distributed systems were introduced that performed amplification, filtering, digitization, and multiplexing at or near the receiver stations. GeoSystems and Opseis were two early market leaders. By the mid-1980s distributed systems were in wide use throughout the industry.

In land recording using a nondistributed system, an analog seismic signal travels from the geophones along electrical conductors (the cable) to a roll-along switch in the recording truck (or "doghouse" or "dog-box"), after which it is converted to a digital signal and recorded on tape or disk. In contrast, in a distributed system, the seismic signal passes from the geophone string directly into an amplifier and/or A/D converter, after which it travels in digital form along a cable to the recording truck. Because digital transmission of multiplexed data uses many fewer cables than analog transmission, layout of the large receiver spreads often used for 3-D acquisition became considerably simpler.

4.2.1 Basic Components

The basic components of the land recording system are reviewed next, using the generic schematic shown in Figure 122. This review is general, since each actual recording system has unique aspects that can be understood best by studying the manufacturer's documentation. The discussion also applies to distributed systems if one imagines that the main amplifier, gain controller, and A/D converter are positioned before the multiplexer.

The components are as follows:

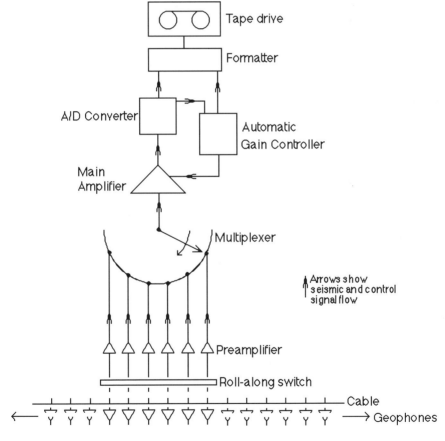

Fig. 122. Nondistributed land recording system schematic.

Roll-along switch—Used in land surveying and located in the recording truck, it allows the observer to record a selected subset of the geophones connected to the recording truck. The roll-along switch minimizes the need to move the recording truck. The switch has many input seismic channels and outputs the seismic data to the recording system. In marine streamer recording, a roll-along switch is unnecessary because every hydrophone group is recorded continually by a dedicated instrument channel.

Preamplifier—This is a fixed gain amplifier that raises the incoming seismic signal above the background instrument noise level. The preamplifier has low noise, high input impedance and low distortion. Its input impedance is equal to or greater than the cable impedance to the furthest station so that no signal amplitude is lost because of mismatching of impedances. The amplifier must be completely linear over its operating range.

Multiplexer—This is an electronic switch that time shares data from multiple channels. It changes multiple parallel inputs to a serial output ready for amplification, digitizing, and recording. The multiplexer cycles through all of the inputs during each digital sampling interval.

Main amplifier—This amplifier receives all analog signals input to it and passes them on to the A/D converter with an amount of gain determined by the gain controller.

A/D converter—Analog signals are converted to digital signals with this device. It allows the analog stream of data to be recorded in digital form. The received incoming signal must be filtered to prevent aliasing prior to conversion to a digital form. This is discussed in more detail below.

Gain controller—The received signal includes reflections, refractions, ground roll, and environmental noise, all of which may have amplitudes varying in a range from microvolts to volts. A fixed form of amplification with only a relatively small number of data bits (such as 14) cannot handle that range without some clipping at the most significant bit end of the converter. Instead, a variable or automatic gain control (AGC) level is determined for application by the main amplifier in the feedback loop with the A/D converter to reduce or amplify incoming signal to keep signal levels within the desired converter range. The controller sets the amount of gain while the amplifier applies it to the incoming signal. The AGC level set at each sample is recorded on tape as part of the gain word.

Formatter—Next, the formatter puts the data stream (in the form of voltage and gain levels) into a binary code for writing onto magnetic tape. In addition, instrument operational commands are distributed by the formatter to all the other components, making the formatter the "brain" of the recording operation.

Tape drive—Data finally are recorded on tape in digital form, ready to be passed on to the processing center for further processing. Magnetic tape may be replaced by floppy disks, depending upon the system in use.

4.2.2 Instrument Noise and Sampling

A major component of all seismic recording instruments, which is not shown in Figure 122, is the recording filter system. The received seismic signal must undergo frequency filtering prior to the recording process to ensure that it can be reproduced later in the data processing center with no confusion of its frequency content (known as *aliasing*). Recording-system filters also are used to remove unwanted noise. The issues of noise and aliasing must be fully understood before a discussion of instrument filtering is possible.

4.2.2.1 Instrument Noise

Several types of electronic noise can occur in recording instruments. Coherent repetitive signals (50, 60, or 100 Hz) are caused by electrical interference or *crossfeed* between the instruments and nearby power supplies or cables. Such noise can be attenuated by shielding or rerouting cables. Modern electronic instruments can suffer from several types of incoherent noise that are inherent in the nature of the components. For example, *Johnson noise* (or *thermal noise*) is caused by the randomness of conducting electrons, whereas *shot noise* is caused by the discreteness of charge carriers within semiconductors. *Quantization noise* is another type of incoherent noise that is caused by the finite width of the voltage level represented by the least significant bit in a digitized signal. Instrument designers strive to minimize these kinds of noise through various design strategies.

If the instrument noise is outside the desired seismic signal spectrum, analog or digital filtering may remove it. If the noise is coherent and of a frequency similar to the seismic signal, then it is difficult to remove without also removing some seismic signal. Incoherent spiking transients may be removed by special data processing techniques performed in the processing center.

4.2.2.2 Sampling

As stated earlier, analog signals must be sampled correctly when digitizing. The seismic signal is sampled at discrete time intervals called the *sample intervals*, which can vary from 0.25 ms to 8 ms depending on the instruments in use (industry often uses the term "sample rate" to mean sample interval). The smaller the sample interval, the larger the potential frequency range that can be recorded. The choice of sample rate is dependent on the seismic objectives and the required bandwidth. As was explained in Chapter 2, to digitize an analog signal properly, it must be sampled at least twice per cycle of the

highest frequency present in the signal. This is called the *Nyquist criterion*. For
a given sample interval, the highest frequency that satisfies this criterion is
called the *Nyquist frequency*. Frequencies above the Nyquist frequency are
sampled incorrectly and are said to be aliased.

Consider Figure 123, in which an 80 Hz sinusoidal wave has been sampled
at an interval of 4 ms. The wave's amplitude values are shown, relative to a
peak value of 1, and the sample times are shown at 4 ms, 8 ms, 12 ms, and so
on. If a 170-Hz wave is sampled now at the same interval, as shown by the
second wave, we can see that the values are exactly the same as for the 80-Hz
wave, so sampling and reconstructing a 170-Hz wave at a 4-ms sample rate
will produce an 80-Hz wave. Hence, a 170-Hz wave is an alias of an 80-Hz
wave. This example shows that a 4-ms sample interval is too low to sample a
170-Hz wave adequately. Taking the example to its extremes, if we sample a

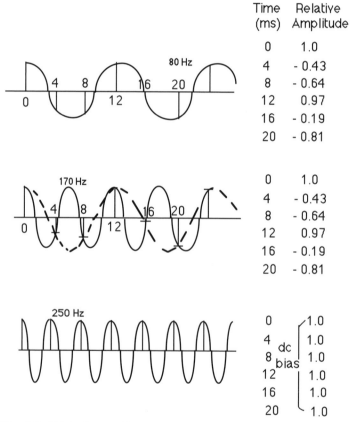

Time (ms)	Relative Amplitude
0	1.0
4	- 0.43
8	- 0.64
12	0.97
16	- 0.19
20	- 0.81
0	1.0
4	- 0.43
8	- 0.64
12	0.97
16	- 0.19
20	- 0.81
0	1.0
4	1.0 (dc
8	1.0 bias)
12	1.0
16	1.0
20	1.0

Fig. 123. Digital sampling.

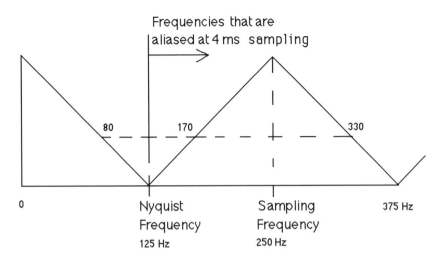

Fig. 124. Frequency aliasing at 4 ms.

250-Hz wave (the frequency for a period of 4 ms), a constant output value will result (called a *dc bias*).

Since aliasing commences at the Nyquist frequency, and the Nyquist frequency is half the sampling frequency, for 4-ms sampling, all frequencies above 125 Hz are aliased. For 2-ms sampling, aliasing starts at 250 Hz; for 1-ms sampling, aliasing starts at 500 Hz. One major role of filters in seismic instruments is to remove frequencies that will be aliased if they are present when the signal is digitally sampled.

Figure 124 shows a method to work out the value of an alias at a particular sample interval. First, triangles are constructed to each side of the Nyquist frequency. Then a horizontal line is drawn across the figure. Aliased frequencies at the intersection of the line with the second triangle will appear, when sampled, to be the frequency at the intersection of the line with the first triangle. For example, with a 4-ms sampling interval, analog frequencies of 170 Hz and 330 Hz will be represented as 80 Hz after digitization.

4.3 Filtering

Figure 125a shows an amplitude response plot (see Appendix) of an idealized seismic instrument filter. The graph is plotted in logarithmic form, with the peak signal amplitude at a value of 0 dB. The frequency range over which signal is allowed to pass in such a filter is known as the *passband,* and a filter which has such a passband is called a *band-pass filter.* The frequency range over which a filter attenuates signal is called the *reject area* or reject zone.

The passband of the filter of Figure 125 is conventionally defined as the frequency interval between –3 dB points (70% amplitude, known as the *half-power point*). The filter *slope* is measured in dB/octave, an octave being the frequency range from a starting frequency to twice that frequency (e.g., 10–20 Hz, 15–30 Hz).

All passband filters cause some amount of amplitude and phase distortion of the true signal. To minimize this distortion and still ensure the filter oper-

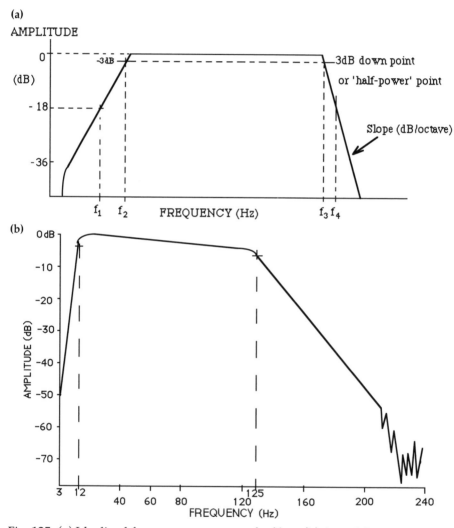

Fig. 125. (a) Idealized frequency response of a filter. (b) Actual frequency response of a 3–125-Hz filter.

ates adequately, it is necessary to use a gentle gradient of the filter slope at the low frequency end (e.g., about 12–26 dB/octave). The slope gradient at the higher frequency end of the spectrum can be 70 dB/octave or more without causing undue high-frequency distortion. The high-end filter is known as the *hi-cut filter* (because it cuts out higher frequencies) or *antialias filter* (because it guards against aliasing), whereas the low-end filter is known as the *lo-cut filter*.

In Figure 125a, if filter points f_1 and f_2 were an octave apart, then the slope would be 15 dB per octave. The passband is between half-power points f_2 and f_3. In practice, analog filters respond with rounded corners, as shown in Figure 125b. We say that the filter response *rolls on* at the lo-cut end and *rolls off* at the hi-cut end of the passband.

In recording or data processing documentation, a filter's frequency range often is expressed solely in frequency terms starting at its lowest operational point, followed by the passband values, and finishing at an end-frequency value. The values for Figure 125b would be 3/12/125/220 Hz (where the effective lowest and highest frequencies to be passed by the filter would be greater than 50 dB down). An alternative approach is the filter-value/slope notation, which expresses the filter in terms of lo-cut/slope followed by the hi-cut/slope value. For example, a lo-cut filter reaching the half-power point at 12 Hz, having a slope of 26 dB/octave from its start up to that point, will be documented as 12/26; a hi-cut filter with a half-power point at 125 Hz and thereafter with a slope of 70 dB/octave will be documented as 125/70. This filter then is written as 12/26/125/70. Documentation must state the filter values clearly in order to avoid confusion during data processing. When the lo-cut filter is switched off so that no low frequencies are attenuated, the lo-cut filter is said to be "OUT." Such a filter is sometimes referred to as either a *band-stop filter* or a *low-pass filter*.

The hi-cut filter in a recording instrument is designed to attenuate aliased frequencies so that they will not distort the received signal after it has been digitized. As shown in Figure 126, a given degree of attenuation at the onset of aliasing (125 Hz in this example) can be obtained by dissimilar filter parameters. The 64/12 filter achieves about the same attenuation at 125 Hz as does the 90/70 filter. Beyond 125 Hz, the steeper filter has better rejection. Whether that is important depends on the amplitude of the aliased frequencies in the received signal. Steeper antialias hi-cut filters usually are preferred, however, because they provide a wider recorded signal bandwidth and, hence, potentially better seismic resolution.

Analog hi-cut filters are typically only effective with slopes up to about 70 dB/octave. Steeper slopes are possible, but such filters have undesirable ripples (known as Gibbs's effect) in their time-domain responses. The intro-

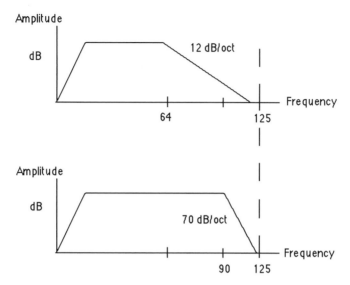

Fig. 126. Use of different filter slopes.

duction of state-of-the-art, 24-bit systems using digital filters has allowed steep slopes of as much as 200 dB/octave to be used, moving the high-cut end of the spectrum close to the Nyquist frequency.

Just as the hi-cut filter value and slope can be changed to alter the upper end of the signal spectrum, so the lo-cut filter value and slope can be changed as desired. It was common practice for many years to use the lo-cut filter to attenuate low-frequency noise when it became a problem. For example, when strong ground roll was observed on land shot records, one approach to removing it was to increase the lo-cut filter value to a point where it attenuated the coherent wavetrains. This approach also has worked in marine recording, where strong streamer strumming (commonly referred to as "streamer jerk") as a result of the vessel or tail buoy moving through rough sea can cause a high-amplitude coherent noise train along the streamer.

More recently, the application of such lo-cut filtering has been discouraged. Increasing the low-frequency cutoff point reduces the recording bandwidth and may cause a loss of useful signal in addition to noise attenuation. Such coherent noise may not have stable characteristics throughout a survey; the reduction in recording bandwidth then becomes unjustified. In addition, different filters and instruments have different phase characteristics so that if different recording filters and instruments are used for a period of years over a survey area, it may become difficult to tie in different vintages of stacked data. Consequently, once a band-pass filter value has been set for an area during an initial survey, ideally that filter should be maintained thereafter during subsequent surveys, in spite of any low-frequency noise problems. Because of their wide dynamic range (see Section 4.6), modern systems can afford to simply record any low-frequency noise and let it be dealt with during data processing.

In land seismic surveying, when power lines are crossing above the seismic cable, the cable conductors may pick up electrical noise at the power-line frequency. Such electrical pickup often can be so high in amplitude that it completely masks the useful seismic signal. One way to avoid this problem is to use a recording filter to reject the power-line frequency. A filter designed to remove a single frequency is known as a *notch filter*. The notch may be at 50 or 60 Hz, depending on the local power-line frequency.

A notch filter can have an undesirable phase response. Using the notation of frequency-value/slope, Figure 127 shows the filter response of a DFS-V recording instrument using a sample rate of 2 ms and having 12/26/124/70 Hz filters (where the passband is 12 to 124 Hz). Such spectra are obtained by Fourier analysis of the instrument impulse response. That is, an electrical spike is input into the recording system, recorded, and then transformed into the frequency domain. The amplitude response in this case indicates that a 60-Hz notch filter is in use. The phase spectrum shows the instrument response with and without the 60-Hz notch filter. Clearly, the use of the notch filter makes a major change to the filter response, causing a phase reversal (i.e., 180° phase change) at frequencies of about 60 Hz.

Figure 127 also shows how a notch filter changes the time-domain representation of an instrument impulse response. Clearly, use of a notch filter in a recording system makes its impulse response less desirable from an interpreter's viewpoint. Another problem with a notch filter is that it not only removes the noise at the notch frequency but also the signal at that frequency. Consequently, the use of such a notch filter in recording instruments is discouraged; other field approaches (such as applying an input signal to the affected channels of exactly the same frequency and reversed phase as the power line) can resolve the problem of power-line pickup in the field.

Figure 128 shows the temporal impulse response of a DFS-V instrument using different filters and how the response would appear after transforming each filter to its zero-phase equivalent (ZPE). The ripples after the initial spike show the effect of the notch filter. Note how the ZPE wavelet has more ripples as the filter passband reduces. This is not a desirable effect because when seismic data are stacked to produce a seismic section, such ripples can sum to produce either a *ringing wavelet* or a lower stack frequency content, thereby reducing horizon resolution.

If the optimum response for a recording instrument is a spike, then the best response in Figure 128 is the OUT-124 Hz option, which most closely resembles an input spike. This is logical because, of these filters, the OUT-124

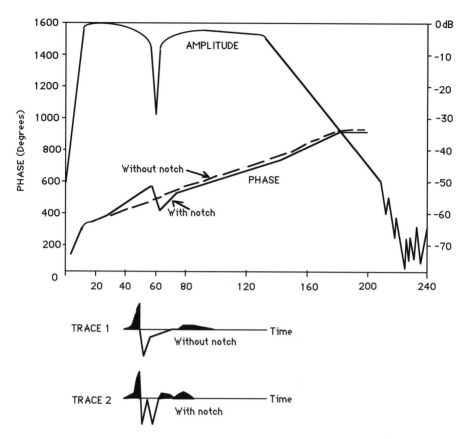

Fig. 127. Amplitude and phase response of a DFS-V instrument, with and without a notch filter.

Hz filter applies the least amount of filtering. By comparison, the 60-Hz notch filter causes ringing and phase shifts, as explained earlier. Often such ringing and phase shifts are removed from the seismic data by an inverse filtering operation called *designature*. This helps in making the seismic sections easier to interpret.

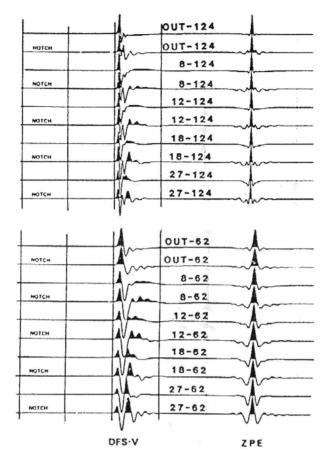

Fig. 128. DFS-V instrument impulse response using different filters.

4.4 Amplification

Received seismic signals have a wide dynamic range; that is, at times the signals are quite large, whereas at other times the signals are very small. On the other hand, recording devices, whether analog or digital, have limited dynamic ranges. When a recording device is presented with a signal outside of its dynamic range, that signal cannot be recorded faithfully. Signals that are too small are not recorded at all, and those that are too large are distorted or "clipped." Seismic recording systems contain an amplifier, the role of which is to match the amplitude of the received signal to the dynamic range of the recording device so that the signal can be recorded faithfully. This amplitude matching is called *gain control.*

In seismic recording instruments, gain control can be either constant or "fixed," dynamic or "ranging," or a combination of both. In a constant gain system, the amplification is set to a fixed amount so that the highest amplitude of interest in the received signal is recorded without being clipped. The amount of fixed gain can be determined by performing and analyzing *gain-constant test shots* at the start of a survey. In a fixed-gain system, small signals are lost if their voltage level falls below the recordable threshold. For example, if an A/D converter (see next section) represents each digital sample by 12 bits and the largest signal is amplified to the point that it is represented by all 12 bits being "on," then the smallest recordable signal is represented by just one bit being on. Such a system has a recordable dynamic range of 72 dB because each bit represents a factor of 2, or 6 dB.

The limitations of fixed-gain amplification can be overcome through the use of gain ranging. In a gain-ranging system, the amplification factor changes on a sample-by-sample basis so that the amplified signal falls within the dynamic range of the A/D converter. The gain for each sample must be recorded so that true amplitude levels can be recovered when the data are processed. The so-called *instantaneous floating point* (IFP) systems introduced in the 1960s used a 14-bit data word and a 4-bit gain word, for a total dynamic range of about 108 dB. Because of this increased dynamic range, IFP systems were adopted industry-wide soon after their introduction.

In the 1990s A/D converters became available that had 24 bits of dynamic range. In these systems, gain ranging is unnecessary, so, once again, fixed-gain amplification is in vogue.

4.5   A/D Conversion

The conversion process is performed by the analog-to-digital converter. The A/D converter samples the input seismic analog data amplitude at discrete points in time and represents each sample by a digital *word*. The greater the number of bits in each digital word, the greater the range of amplitude that can be converted without distortion. If a recording instrument has 14 bits, it can convert signal up to a value of 2^{14} or 16 384. Therefore, 14 bits plus a sign bit can represent an amplitude from $-16\,384$ to $+16\,384$.

The smallest signal that can be converted is that signal value represented by the least significant bit (lsb) of the converter. The dynamic range in the above example is then 16 384 to 1 or 84 dB (see Appendix A for this relationship). In practice, the absolute converter range is slightly less since the amplifier has an ambient background noise value which is converted into a value in bits. For example, in some IFP amplifiers, the lsb is used up by converting the instrument noise alone, reducing the total system dynamic range by 6 dB.

An amplifier has ambient noise, whereas a converter does not. If it were possible to increase the number of bits that could be converted, then the dynamic-gain amplifier could be replaced by a fixed-gain amplifier (in the form of the GC mentioned earlier), thus reducing amplifier noise. A 24-bit converter expands the dynamic range to a theoretical value of 138 dB. Instrument manufacturers have introduced this innovation into their recording systems, which because of internal system noise have a practical dynamic of 114 dB. The main land manufacturers are Input/Output Inc. and OYO Inc. Input/Output has a 24-bit system widely used in the marine surveying industry, and Syntron Inc. also is marketing similar systems.

These systems can have advantages over the previous dynamic gain-ranging systems, apart from the ability to reduce amplifier noise. For example, a major advantage of the Input/Output system is that the antialias filter is digital rather than analog. Consequently, the antialias filter is zero-phase, so it does not change the phase of the recorded signal. Digital filtering also allows a broadening of the passband spectrum up to 3/4 Nyquist frequency with a steeper 200-dB/octave filter slope, rather than the 1/2 Nyquist frequency and 70 dB/octave slope required by the analog filters (see earlier in this chapter). Most 24-bit systems have such filters after digitization so that different filters can be used without affecting the input data. If filters are not available to attenuate ground roll prior to A/D conversion on a system (they were available with the previous AGC systems), then this would be a drawback of such systems. A schematic of one type of A/D converter is shown in Figure 129. The output of the A/D converter is a digital representation of a voltage sample, known as the *seismic word*. When written on magnetic tape, a seismic word represents a single seismic sample—one voltage value. A seismic word has three parts—a data word (often referred to mathematically as the *mantissa*), a gain word (referred to as the *exponent* or *characteristic*), and a sign bit. A data word is the basic value that has been converted, whereas the gain word represents the amount of gain to be applied to the data word to restore it to its true analog value.

In practice, the signal is connected to a sample and hold circuit for approximately 1 µs. This charges a capacitor to a voltage which remains and is used for comparison for 30 µs with a series of fixed-ladder voltages. When the appropriate voltage comparison is made, this value is supplied at the next sample time. The voltage is on a binary scale. If the voltage level has attained a maximum value, the AGC is requested to step down 12 dB on the next sample. If a minimum converter value has been reached, the AGC steps up 12 dB. This range of gain is known as the *gain window*. The output is then sent with all logical commands to the formatter, which puts the bits into a format for

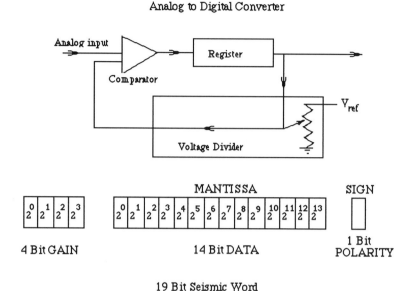

Fig. 129. Analog-to-digital converter schematic and gain word.

writing magnetically on tape. A full gain of 84 dB may be stepped up or down in one sample.

In the decimal number system, 10 is the base number. All numbers can be represented as a function of 10. For example, a decimal system data word that has a value of one thousand is represented by 1000. In the binary system, 2 is the base number, where a data word is represented by a series of powers of 2. A binary data word which has a value of 1 is then represented by 2^0; a data word with a value of 2 is represented by 2^1; 2^2 represents 4; 2^3 represents 8, and so on. So, if a data word has the values $2^0 + 2^1 + 2^2$, it represents a value of $1 + 2 + 4 = 7$. If the gain word has a value of 2^1, the data word value of 7 should be multiplied by 2 to produce the true value of 14. To take this example to its conclusion, a true value of +14 will have a seismic word consisting of a gain word of 2^1, followed by a data word of $2^0 + 2^1 + 2^2$, and a positive sign bit.

4.5.1 Converter Operation

In Figure 129, consider what happens when a seismic voltage sample arrives at the converter. A value of voltage is already resident on the first unit with which it comes into contact—the *comparator unit*. The two values of voltage are compared, and if they are the same, the register outputs a signal to the logic circuitry to record a binary word containing a number of data bits and a sign bit representing the comparator voltage. A gain value of 1 (or 2^0) in the gain word would be used to maintain the same data value to be written to magnetic tape. When the next sample arrives at the comparator—let us assume it is a quarter of the previous sample value—then the comparator will inform the register that it has a lower voltage, and the register informs the voltage divider that it must reduce its value. The divider reduces by one bit (i.e., divides the reference voltage V_{ref} by 4), and this value is then input to the comparator, and the gain word becomes 2^2 (i.e., 12 dB). The two inputs at the comparator are now the same again and the new binary word representation (consisting of a data word and a gain word 2^2) of the voltage sample is written to tape. Any mantissa bit up to 2^{13} may be stepped in one sample, which is 84 dB as discussed previously.

The lowest sample amount that can be converted is typically 0.25 μV, so that an A/D converter with a total of 120 dB dynamic range would be able to digitize signal in a range from 0.25 μV to 0.262144 V (i.e., 2^{20}). Allowing for the sign bit, the most significant bit represents ± 0.262144 V.

4.6 Dynamic Range

Instrument dynamic range (DR) is the ratio of the largest storable signal on the system to the smallest background noise. The storable signal is the amount of signal amplitude that can be digitized and recorded on a medium (such as magnetic tape) and subsequently recovered in the processing center without distortion. DR is influenced by the instrument system noise as a result of the chosen gain constant value and the number of converter bits available to represent the analog signal in a digital form. To explain this concept, we will use, as an example, a recording system having both fixed and dynamic gain control.

The total instrument gain is the amount of fixed gain available (the GC) in the instrument plus the amount of variable gain which can be applied by the amplifier to record the signal without distortion. This variable amplifier gain is applied to the signal in steps of 6 or 12 dB. All incoming signals have the initial fixed GC gain applied to them (which may have a maximum value of 30 dB), followed by an amount of variable amplifier gain, depending on the amount of gain required to correctly record the signals. The variable amplifier gain normally can apply a maximum amount of 90 dB to the smallest signals. Therefore, the total amount of gain possible is the fixed GC plus the variable 90 dB, or 120 dB.

In the gain schematic of Figure 130, if the system noise level is 120 dB less than the signal that can be recorded without applying any gain, and the

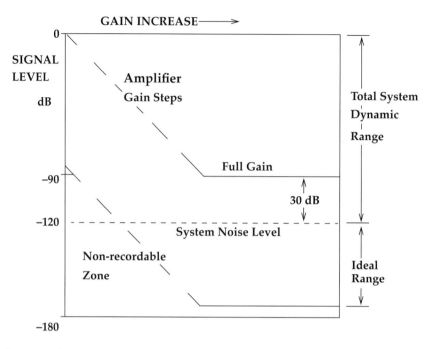

Fig. 130. System dynamic range as a function of gain.

GC plus the amplifier gain is 90 dB, then the remaining DR available will be only 30 dB. Conversely, if the GC is set at 24 dB, then the DR available will be 120 − 24 = 96 dB above the system noise. The maximum amount of gain available in any instrument is limited by the number of analog-to-digital conversion bits available in the instrument converter. We ideally would like an infinite DR in the recording instruments, but this is not necessarily useful since eventually low-level signals become completely lost in instrument noise.

4.7 Recording

4.7.1 Formats

Because seismic data may not be processed by the same company that recorded them, data must be recorded in a standard field format that can be read by any processing center. For this reason, the Society of Exploration Geophysicists (SEG) established a series of standard tape formats for the exploration industry (SEG, 1980). As a result, data in the past have been recorded in

SEG B, C, or D formats on magnetic tape. Generally in seismic recording, data are written in the SEG D format, be the tape in nine-track, DAT, cartridge or Exabyte form. This is usually transcribed in the processing center to the SEG Y format—the industry standard format which all processing centers can read.

The field formats differ in the number of channels that can be recorded and the amount of space available to record peripheral data (such as filter settings). Such space is called *header space,* so that peripheral data in a header are often called *header information.* Before the early 1970s, recording systems used up to 21-track tape (1-inch width). Today, nine-track tape is the most common. Use of cassette-type cartridge tape is prevalent in marine recording. The SEG D format is shown as an example in Figure 131.

Data bits on magnetic tape are written and read by the so-called *write head* and *read head,* respectively. As a quality-control check to ensure proper data recording, the read-after-write head reads the data immediately after they are recorded. The tape drive (or *tape transport*) read head is connected to a playback unit, which allows an immediate replay of data. This unit also may be connected to an oscilloscope, a video display terminal, or a *camera* unit; these allow a visual check of recorded shot data.

The seismic camera is a wiggle-trace plotter, which plots the wiggle traces on paper as a *camera monitor record.* Alternatively, an electrostatic plotter (e.g., Versatec) or a thermal plotter (e.g., OYO) may perform this duty. The camera monitor received its name historically; in the early days of seismic recording, the only form of record of the wiggle traces (before the invention of the magnetic tape drive) was a photographic camera. A voltage output representing a seismic trace from the recording unit was fed to a galvanometer, from which was suspended a mirror. As the voltage moved positive, then negative, the mirror rotated. A light beam reflecting off the mirror was thus deflected across a piece of photographic paper, and this represented the seismic trace. In those days, seismic crews also had a darkroom alongside the recording room.

Dual tape drives are used in surveys when high data production rates leave insufficient time to rewind and replace a full tape reel between shots. Where cartridge tape drives are used, a number of cartridge drives may operate at any one time, depending on how many data channels are being recorded. Tape speed is quoted in inches per second (ips) and may be checked by an oscilloscope, a hand gauge, or an LED display. Tape speed should not vary more than ±2%, although data have been recovered from tapes recorded with a tape speed error as high as 15%. Loss of tape-speed timing synchronization is the most common cause of unreadable tapes.

The volume of data on a tape depends on the tape's bit-packing density. A 6250-bits-per-inch (bpi) density system will pack nearly four times the data as

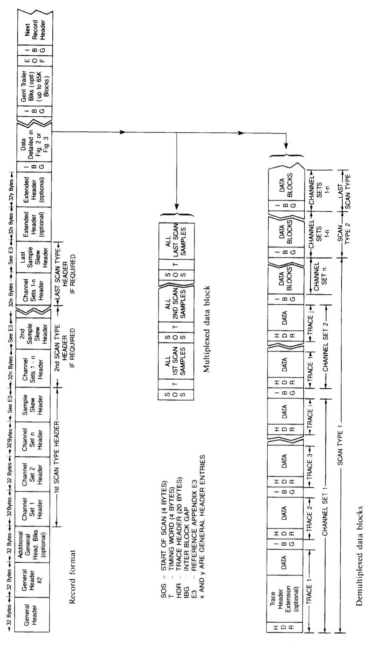

Fig. 131. SEG D tape format.

a system that records at a 1600-bpi packing density. The packing density of magnetic tape is given by

$$\text{Packing density BPI} = \frac{\text{Data rate}}{\text{Tape speed}}. \tag{36}$$

Cartridge tapes can record data at 32 000 bpi, which is a substantial increase in data per unit area compared with the older systems, which recorded data at 800, 1200, 1600, and 3200 bpi. Cartridge tapes that have higher packing densities are expected in the near future.

4.7.2 Recording Channels

Recording instruments such as the System Two (Input/Output Inc. trade name) can record as many as 4000 channels of data in SEG D format at 6250 bpi. In marine recording, especially 3-D recording, data often are recorded from three or more seismic streamers, each of 240 channels, at one time. Hence, it is realistic to record with at least 720 channels in marine surveys, although single- and double-streamer operations (240 and 480 channels, respectively) are more common in conventional marine 2-D recording. Land surveying often is performed using 120 to 240 channels. Sometimes, an increase in the number of channels can cause limitations elsewhere within the recording instruments. For example, a set of instruments may be capable of recording 240 channels at a 2-ms sample rate, but to record 480 channels, a sample interval of 4 ms is required. Here, there is the question of which is more important, the maximum number of channels that may be used (thereby possibly increasing the fold) or the faster 2-ms sample rate (thereby doubling the recording frequency range).

Often, by using more equipment, more channels can be recorded. The economics of equipment availability then becomes an important part of the limitations compromise. In today's complicated 3-D surveys, firm conclusions regarding the optimum number of channels are difficult to reach. That is why the industry has such a proliferation of survey designs.

The Geocor IV, marketed by GeoSystems, was the leader in channel expansion technology in the early 1980s. It could record 1024 channels when other instruments could record only 120 channels. It did this by the use of sign-bit recording, as opposed to full-word recording (see later). Two advantages of this system were that it could perform real-time vibrator correlation and fast Fourier transforms. However, the increasing sophistication of electronics has caught up with this system; today real-time vibrator correlation and transformations can be performed on full-word instruments recording a high number of channels.

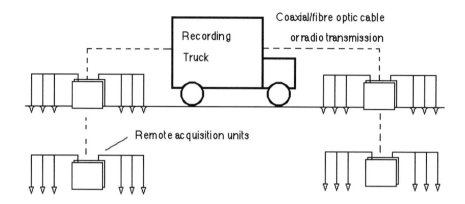

Fig. 132. Distributed systems using land line or telemetry.

Manufacturers such as GUS, Sercel, and Input/Output market distributed systems that utilize twin-wire coaxial cable or fiber-optic cable, connected to individual remote seismic data-acquisition units. Each station acquisition unit can take as many as six stations of geophones, and their output is transmitted in serial format to the recording truck (Figure 132). It is a simple and fast operation to lay a coaxial or fiber-optic cable down on the ground and plug into it an acquisition unit that receives, amplifies, samples, and multiplexes six stations of data. Such field units simplify field work while allowing the use of a greater number of recording channels.

4.8 Miscellaneous

4.8.1 Telemetry

Telemetry is the act of transmitting signals from remote locations to a central receiving point, either by cable (coaxial or fiber optics) or by radio. Each remote acquisition unit typically consists of a preamplifier (fixed or variable gain), a converter for digitization, and a data transmitter. In the case of cable transmission, data are sent in packets from one remote unit to another, cascading into a stream, so that a large number of remote stations eventually may feed data to a single receiver in the recording truck. With radio operation, the recording truck has individual receivers which demodulate the radio signals before recording them on tape or disk.

Because radio telemetry systems rely on radio transmission of the seismic signals, they may perform poorly in mountainous terrain due to a loss in radio signal strength. However, radio telemetry has been successfully used in remote areas over shallow water, on mud and tidal flats, and in swamps and jungles.

The advantages of telemetry over conventional land systems are:

1) Digital conversion at each station allows an increase in the number of recording channels that may be used.

2) It is sometimes faster to lay out and pick up the receiver spread because individual personnel can pick up the receiver units rather than wait for a cable truck.

3) The ability to position remote receiver units anywhere allows more flexibility in multiline receiver spread geometry.

4) Because there are no lengthy cable pairs running to a recording truck, there are no power-line noise pickup problems. Power-line pickup does not occur with fiber optic cables, either.

5) Data parity checks can be performed before transmission and after reception to ensure that data have been transmitted and recorded correctly.

6) Individual stations and remote units can be transported by helicopter faster than other types of cables.

7) Telemetry units can avoid cable-laying obstacles such as rivers, railways, and highways.

Apart from the poor radio transmission, another problem with the use of any remote systems is that they obtain their power from batteries that require frequent recharging. Solar panels can charge unit batteries during the day, but such energy is dependent on the sun shining, so the batteries still must be checked frequently to ensure no failures occur.

Texaco was an original user of telemetry transmission in the 1960s. Aquatronics developed Teleseis, a system that replaces cables with a radio link, in the 1970s. A telemetry leader in the 1980s was Applied Automation's Opseis system, which scanned more than 1000 channels at 2 ms. The system used a cassette recorder at each station to record data to tape after each shot had been fired. The tapes were retrieved from each acquisition unit at the end of each day.

Where the survey area is flat with few obstacles to recording operations, most crews tend to use the coaxial or fiber optic cable approach. However, where there are ravines or hazards to the smooth flow of operations, radio telemetry often is preferred, weather permitting.

4.8.2 Sign-Bit Recording

Sign-bit recording came into use in the early 1980s. This technique, originally devised by NASA as a method of enhancing low signal-to-noise ratio transmissions from a Venus space probe, uses vertical stacking of successive records to enhance the signal-to-noise. This method later was applied to recording seismic traces with great success.

Vertical stacking to enhance the signal works well, provided the noise amplitude level is equal to or less than that of the signal. But if the noise is recorded in random bursts with an amplitude value much higher than the signal, it will take many stacks of records to reduce the noise bursts to the signal level. This problem is overcome if all values of signal and noise are limited to plus or minus 1. In this case, strong or weak signals have a value of ±1 and are always enhanced by the vertical stacking process. By comparison, large-amplitude random noise bursts are limited to ±1, so that stacking further shots reduces the noise level.

A system that records a value of +1 or −1 for any positive or negative signal is known as a *sign-bit recording system,* since it only need record the sign of the amplitude, the value itself always being 1. A value of zero is not recorded, and the converter only needs to know whether the signal is positive or negative.

Figure 133 shows how sign-bit recording can be equivalent to an 18-bit recording of a seismic trace. Random noise causes the zero crossing of the input trace to shift from record to record; after 16 stacks (sums), the sign-bit recording is almost identical to the conventional 18-bit recording. This only works if there is random noise. If the noise level is too low, then additional noise must be generated (this is very easy when recording a land seismic line because line vehicles are constantly moving along the receiver line). Sign-bit recording therefore works well under high-noise conditions, in which quality is improved by recording many apparently poor-quality (low signal-to-noise-ratio) records.

Where high-amplitude random noise bursts occur and the signal-to-noise ratio is less than 1:1, the sign-bit technique is effective. This may prove beneficial, for example, when working in populated areas having random traffic-generating noise or when working in seismically poor areas of weak reflections and occasional sideswipe.

The sign-bit technique is marketed by GeoSystems Corporation (which, although it has become insolvent in the United States, still is active elsewhere) as its petroleum survey Geocor instruments, and by Sercel as its Mini-Sosie high-resolution shallow survey instruments. The Mini-Sosie, rather than producing a programmed sweep, uses a random sweep, but apart from that aspect, its recording system is similar in operation to the Geocor system. The Mini-Sosie generates its own random noise bursts as the rammers are manually pushed along the seismic line. Geocor and Mini-Sosie are the only sign-bit systems currently used by the industry.

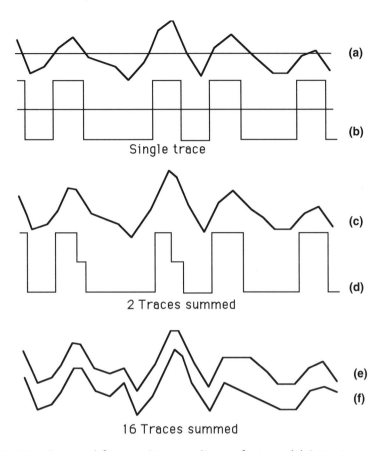

Single trace

2 Traces summed

16 Traces summed

Fig. 133. Simulation of the sign-bit recording technique. (a) Seismic trace recorded with a full-word (18-bit) instrument. (b) Data recorded with sign bit only. (c) Second recording and sum by a full-word instrument. (d) Sign-bit data after two sums. (e) The conventional full-word record after summing 16 times. (f) The sign-bit equivalent after 16 sums. In the sign-bit case, additional random noise is added to the trace prior to summation.

4.8.3 Field Computers

Computers in the field can provide powerful support for various aspects of the recording operation. A vibroseis uncorrelated field record is uninterpretable, so this computation always has been performed in the field. Early field data processing computers were based on the DEC PDP-11 machine. However, the large size of such machines restricted their common use until the advent of the IBM personal computer (PC). The PC reduced the cost of processing but was frequently too slow or had inadequate software to perform much more than the simplest of input/output functions. UNIX-based workstations were then developed to be more powerful than the PCs. Today, many field crews have data processing workstations to provide quick-look and general data processing support during field acquisition.

Interactive field computers are considered necessary during land crew startup when test lines and source tests need to be evaluated. Processing costs and time at the computer center can be saved using a field computer system that can demultiplex field records. Three-dimensional data acquisition, both land and marine, would be almost impossible today without some form of field computing—even if it were only to locate the position of common midpoints during the recording operations—to ensure that the fold of coverage is adequate and within specified tolerances.

Field computers have blossomed on marine vessels during the upsurge in 3-D data recording. When four streamers are collecting data from four source arrays, the amount of positioning information for recording increases substantially. Networked workstations are becoming the norm for recording and processing the navigation sensor data in near real time. For example, a streamer's depth, feathering angle, and x,y location can be updated every second using ship-monitoring computers. The collected data also can be inspected to ensure that the quality of recording is acceptable.

Many recording systems have computers able to perform on-line phone tests and analyses as well as cable tests prior to each shot. This is useful in checking receiver integrity before recording commences. A number of instruments are able to perform limited signal processing as a *quick-look* data processing package. The advantage of having a system that can do some form of field processing is that interpretation of field stacks may identify interesting formations that could be further delineated by a modified program. One quick-look approach in marine 3-D recording is to bin short offset traces in a low fold 3-D volume, which may be rapidly processed and provide an early indication of data quality as well as profiles and time slices through the 3-D volume.

Exercise 4.1

1) If two seismic lines (which tie at their respective centers) were recorded by different source, receivers, and instruments, what tests would be needed on the field-acquisition system to ensure that the data phase ties in data processing would be made correctly?

2) (a) Surface waves take up a substantial amount of the dynamic range of floating-point recording instrumentation. Why should this be less of a problem when using 24-bit instrumentation?

(b) If the least significant bit of a 96 dB dynamic range system represents 0.25 μV, and a peak reflection of 1 mV is detected at the receivers accompanied by a surface wave valued at 2 V peak, would the dynamic range be adequate to record the reflection?

Chapter 5

Survey Positioning

5.1 Introduction

Accurate positioning of a seismic line is as crucial as having the best possible data quality. Positioning is important for three reasons: (1) many data processing steps require accurate relative source and receiver positions; (2) tying several seismic lines together requires knowledge of where they are relative to one another; and (3) when drilling sites are selected from seismic data they have to be referenced back to an actual location on the Earth's surface. Of these reasons, perhaps the last is most important: No exploration company wants to spend millions of dollars drilling, only to miss the target because the seismic data were mispositioned.

Accurate positioning is not a trivial task, especially for marine surveys. Once a seismic vessel has sailed along an intended line, no permanent evidence remains behind to show where the ship actually sailed. Furthermore, at sea, intended shot and receiver positions cannot be identified by markers prior to shooting. Finally, during shooting, both the ship and the trailing equipment are somewhat at the mercy of the wind, currents, and wave action; the position of the shots and receivers cannot, therefore, be controlled accurately. For these reasons, positioning in marine surveys is a so-called real-time activity; that is, position measurements have to be made, recorded, and processed as a line is shot.

For land seismic surveys, positioning does not have the real-time urgency that it does in marine surveys. The shot and receiver positions can be marked on the ground either before or during the shooting of a line. Likewise, accurate marker positions can be measured leisurely at any time. Furthermore, land surveys have the luxury of being referenced to *permanent markers*, locations that are unlikely ever to be moved. Nevertheless, accurate land positioning is not simple. Survey terrains are not flat, so land positioning must include accurate elevation measurements, a dimension that is not so crucial for marine surveys. In swamps, heavily forested regions, and mountainous

187

areas, accessibility and line-of-sight problems can make land positioning difficult.

Geophysicists refer to locating a seismic line on land as *surveying the line* because the positioning technology is based on traditional land-surveying methods. The hydrographer or *marine surveyor* employed in the marine industry to locate lines refers to the same task as *positioning*. There are other distinct differences between land and marine surveying. For example, in land surveying, distances along seismic lines have been measured for many years using theodolite and compass instruments, whereas in marine positioning, the position of the seismic ship is determined by radio navigation equipment, which requires radio towers to be erected along the shoreline as well as aboard ship.

While these differences have distinguished land from marine surveying in the past, the introduction of GPS satellite positioning systems over the past few years has meant a unifying of positioning methods using a common technology. Today, 3-D survey coordinates can be computed on a small computer as the surveyor walks along the seismic line on land or as the ship cruises along the seismic line offshore. In a few areas of the world, full 24-hour coverage by GPS satellite is not yet available; consequently, the time-tested use of theodolite on land or radio positioning offshore is still prevalent.

Even with satellite control of location, land surveys should be tied into at least one permanent point that has accurately known coordinates. A permanent point, known as a survey *benchmark*, provides a fixed reference for future line and well placement. On regional land surveys with large grids, benchmarks can be located at the ends of each seismic line so that any tie-in errors can be checked. A steel stake (often referred to as a *star picket*) or pipe set in concrete alongside some clearly visible landmark makes a good benchmark. Existing structures such as buildings in populated areas or large trees in remote areas also can be used as survey benchmarks. However, because such structures are not really permanent, a wise surveyor will establish several alternative benchmarks as backups.

In marine surveying, the use of accurate electronic positioning equipment is important because no trace of a survey can be found after a line is shot. Large hydrographic and geophysical research projects typically use navigation instruments with only kilometer accuracy. In contrast, hydrocarbon exploration requires precise positioning to meter accuracy, if possible, because of the limited size of potential oil traps. Consequently, highly accurate radio instruments are used in offshore seismic work. Streamer receivers and sources have transponders (or *pingers*) positioned along their length that transmit and receive signals between each other, from which the relative source and receiver positions can be computed. Cable compasses mounted along the streamer's length measure streamer orientation, data which improve the accuracy of receiver position calculations. Tail buoys often have satellite receivers and transmitters on them, which provide a continuous update of their location with respect to the streamer and towing ship. In surveys in which two ships are used, they are linked by radio, satellite, or acoustics in a network so that the location of all of the equipment is known precisely at any point in time.

Even though survey positioning usually is done by experts in the use of the positioning hardware and software, all data-acquisition geophysicists should understand the fundamentals of this activity. This is especially true for 3-D acquisition, in which the complexity of a survey, the large number of shot and receiver stations, and the need for precise positioning make surveying a crucial and costly part of operation. Viable 3-D surveys cannot be designed without an appreciation of the types and limitations of positioning equipment and an understanding of surveying fundamentals such as spheroidal coordinates, datum corrections, mapping projections, and so on. This chapter will explain the fundamentals of seismic surveying; the more advanced 3-D topics are discussed in Chapter 7.

5.2 Maps and Projections

Seismic surveys usually are planned, processed, and interpreted as if the surface of the Earth were flat. In reality, of course, data acquisition takes place on the irregularly shaped surface of the Earth. To convert survey position coordinates from one type of reference frame to the other can involve one or more complicated mathematical transformations. Further complicating matters is the fact that there is not a universally used set of reference frames. There is a universally recognized coordinate frame representing the Earth's surface called WGS-72, but individual countries and regions tend to prefer local reference frames, mostly for the convenience of having compatibility with historically significant surveys of their areas. Confusion over the usage of different reference frames can spell geophysical disaster. For example, a rig once positioned in the South China Sea using WGS-72 coordinates missed its drilling target by some 500 m because it was not known that the target position was based on a seismic survey that had coordinates given in a local reference frame. This section of Chapter 5 and the next section aim to prevent such blunders by introducing the reader to the various types of reference frames used in survey positioning and explaining how they are related to one another.

The Earth's nearly spherical surface is called the *geoid*. A mathematical approximation of the geoid is called a *spheroid*. The surface of a sphere or a spheroid can be represented (i.e., mapped) as a flat surface in many ways—called *projections*. All of them introduce distortion into the map. Most projections attempt to minimize distortion of a particular type or of a particular region at the expense of increasing distortion elsewhere. Although projections are actually mathematical formulae that explicitly describe how a point on a spheroid is placed onto a flat map, they can be understood easily in terms of visual projections. The most common projection types are conical and cylindrical. A conical projection is obtained by mapping the Earth's spheroid to the surface of a cone. The vertex angle θ is chosen to cause the cone to intersect the sphere at two latitudes (Figure 134). This produces a flat map with low distortion in the middle latitudes. Such a map will have similar distances at the middle latitudes in the east-west direction to the spheroidal distances. However, conical projections do have significant distortion within 30° of the poles and the equator.

A cylindrical projection known as the *transverse mercator*, the type of map most people are familiar with, also has low distortion in the middle latitudes. The Universal Transverse Mercator (UTM) system is widely used by the exploration community. The UTM system divides the Earth into 60 vertical strips or zones, each being 6° wide (Figure 134). The zones are centered at longitudes 3,9,15,..., 357°. These longitudes are called the central meridians (CM) of the zones. The UTM coordinates of a position are called its *northing* and *easting*. By convention, the central meridian of a zone has an easting of 500 000 m. The easting increases for points east of the CM and decreases for

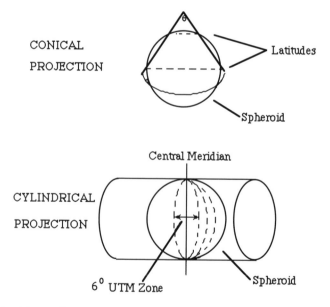

Fig. 134. Different projection types.

points west of the CM (Figure 135). In the northern hemisphere, the equator is assigned a northing value of 0 m. Northing values increase as one moves north of the equator. In the southern hemisphere, the equator is assigned a northing value of 10 000 000 m. As one moves south of the equator, the northing values decrease.

The northings and eastings of a point can be computed in any number of TM projections. For example, point A in Figure 135 could have northings and eastings expressed in reference to either UTM zone 1 or UTM zone 2. Point A also could have northings and eastings expressed, say, in a TM projection with a central meridian passing through point A. In that case, the easting of point A would be 500 000 m exactly. To compare or work with the northings and eastings of two points, they must have a common projection choice. Suppose, for example, that in Figure 135 one has the northing and easting of point B for UTM zone 1 and of point A for UTM zone 2. To compute, say, the distance between the two points, one must either convert both points into latitudes and longitudes or convert one point from one projection to the other. That is, one could find the latitude and longitude of point A and then compute its northing and easting for UTM zone 1. The exploration geophysicist always must be aware that northings and eastings are meaningless unless one knows with which projection they were computed.

Figure 136 shows the various types of projections used for surveying and mapping, and provides an indication of how often they are used and their degree of complexity. The geophysicist tends to deal mainly with TM and UTM projections.

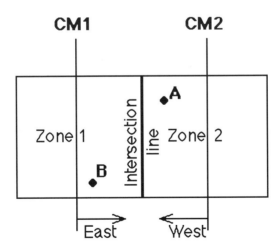

Fig. 135. Intersection between zones.

Projection	Type	Conformal Grids		Arc to chord calculation	Used for surveying	Used for charts
Mercator		Yes	Not often used	Simple	On small scales only	Yes
Transverse Mercator	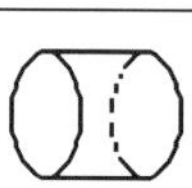	Yes	Universal T.M.	Simple	Yes	No
Conical orthomorphic		Yes	Many with 1 or 2 parallels	Not simple	Yes	A few
Skew orthomorphic		Yes	Malaysia	Simple	Yes	No
Stereographic		Yes	Polar stereographic	Not simple	Yes	No
Gnomonic	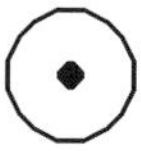	No	Not used	No arc to chord	No	Yes
Cassian	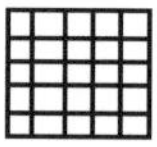	No	Very Many	Difficult	Not recommended	

Fig. 136. Projections used for surveying and mapping (after Admiralty Manual of Hydrographic Surveying, 1965).

5.3   Geodetic Datuming

A spheroid is an oblate ellipsoid of revolution that best fits the form of the area to be surveyed. To describe a spheroid's shape, we refer to the spheroid in terms of its radius of axes. There are two orthogonal directions which determine the shape of an ellipse, known as the semi-major axis, A, and the semi-minor axis, B (Figure 137). A geoid is the best fit surface, which fits the average topography of a local area of the earth to be surveyed and may be centered near but not at sea level for that area. A geoid tends to be perpendicular to the Earth's gravitational field at all points along its length. The topography may vary either side of this local geoid, which may not be far in

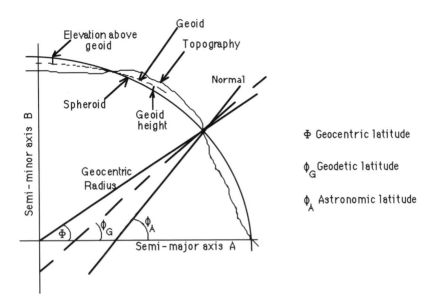

Fig. 137. Geodetic surface relationships.

elevation from the spheroid. Satellites are referenced to a geoidal datum level, not to sea level.

Figure 137 is a schematic figure of the Earth's geodetic surface relationships, which may be determined from a model of the Earth's gravity field. With a knowledge of the geoidal height and a spheroid's axis values, one can determine the differences between a local geoid and spheroid; typical values are shown in Figure 138 for the most common spheroids. The shifts are used to translate the satellite position figures into the local coordinates. This is important when working over large land or sea areas where the Earth's curvature becomes a factor in the survey line positioning accuracy.

The difference between spheroid and geoid depends on the selection of the semi-major A and the semi-minor B axes. In the early surveying days before satellite positioning, there was no need to tie continents together because all surveying was local. Hence, surveyors chose a spheroid to fit local areas. For example, the Everest spheroid has values of A and B to obtain a best fit of the Indian continent to the spheroid produced by those values.

Land surveying uses a best-fit geodetic datum for the area being surveyed. The geodetic datum is defined by the following parameters:

1) A semi-major axis (A)

2) A semi-minor axis (B), or alternatively, flattening (f) where $f = \dfrac{A - B}{A}$

DATUM	SPHEROID	SEMI-MAJOR AXIS	RECIPROCAL FLATTENING	SHIFT TO NWL-8D a = 6378145 1/f = 298.25			SHIFT TO SAO-C7 a = 6378142 1/f = 298.255			SHIFT TO WGS-72 a = 6378135 1/f = 298.26		
		METERS		ΔX	ΔY	ΔZ	ΔX	ΔY	ΔZ	ΔX	ΔY	ΔZ
NAD 1927	CLARKE 1866	6378206	294.98	-23	159	185	-26	155	185	-22*	157*	176*
EUROPEAN	INTERNATIONAL	6378388	297.00	-81	-99	-118	-93	-132	-143	-84	-103	-127
TOKYO	BESSEL	6377397	299.15	-147	530	676	-140	510	689	-140	516	673
INDIAN	EVEREST	6377276	300.80				293	697	228			
AUSTRALIAN NATIONAL	REFERENCE ELLIPSOID 1967	6378160	298.25				-88	-36	86	-122	-41	146
OLD HAWAIIAN MAUI OAHU KAUAI	CLARKE 1866	6378206	294.98	52	-262	-183	59	-263	-203	65 56 46	-272 -268 -271	-197 -187 -181
CAPE (ARC)	CLARKE 1880 (MOD)	6378249	293.47				-130	-147	-347	-129	-131	-282
SOUTH AMERICAN	REFERENCE ELLIPSOID 1967	6378160	298.25				-284	116	-410	-77	3	-45
ARGENTINE	INTERNATIONAL	6378388	297.00				-167	128	25			

Fig. 138. Transformation constants between geodetic systems (after Stansdell, 1978).

3) a latitude and longitude for a point of origin.

A change in any value of (1), (2), or (3) alters the computed values of horizontal and vertical distance along the datum. Hence, there is no conformity in position, distance, or azimuth of surveys which have physical points in common but are based on quite different geodetic data. When comparing locations in two such surveys, a mathematical correction called *datum shifting* is necessary. Datum-shift values often are provided for an area referenced to a central position. Although such shifts are correct for that central position, they are only approximate for any other location. Large areas may require a change to this shift when working some distance from the central position.

For example, satellite receivers in Australia produce positions in the Australian National Spheroid (ANS). These positions (ideally) should be compared with the WGS-72 computation for the area in which the position was computed before the ANS coordinates are accepted as being correct. If the satellite receiver had been used with some other spheroid system prior to mobilization to Australia, it still could be computing the Australian position using the previous spheroidal shift. A vertical error of 4 m and horizontal error of 10 m, could be made. Consequently, before any surveying is performed for a seismic survey, one must ensure that the coordinate system and shift values are correct. Otherwise, gross errors may occur that will affect the final positioning of the lines and of subsequent drilling rigs.

5.4   Land Surveying

Benchmarks are generally located with varying degrees of accuracy. Survey benchmarks can be categorized as first order, second order, or third order. First-order markers, which are positioned by government agencies, are generally of millimeter accuracy and appear in different forms in different countries. For example, in Europe, first-order markers are often a plaque cemented into the ground or a well-known point in a public building, whereas in Australia they may be large steel triangles resident on the top of a hill. Seismic survey benchmarks are often second order, which is of less than 1 m accuracy, and may be recognized as pickets or spikes along a seismic line. (Metal description tags should be attached to pickets rather than paint markings, which wear with age.) Third-order accuracy, which is greater than a meter, is not acceptable for seismic surveying.

In seismic land surveying, accuracies are often less demanding than those required for geodetic and engineering surveys. The amount of positional error acceptable in a seismic survey depends on how the data will be used and the bandwidth of the data. For example, larger source and receiver positional errors may be acceptable in a 2-D survey than those acceptable in a 3-D survey. In general, positional errors are acceptable as long as they do not affect the quality of the final product of the survey. In the past, surveys were conducted by *traversing* from a known benchmark using a survey instrument (theodolite), a rod, and a chain (measuring tape). As a result, short horizontal errors of 1% (10 m per km) often were considered acceptable, while long errors of 0.1% (60 m in 50 km) also were accepted if the survey lines were part of a reconnaissance survey rather than a detailed in-fill survey.

Today, laser distance measurement units (DMUs) with LED displays obtain accuracies better than 0.01% on all line-of-sight ranges up to around 5 km. In hot desert conditions, heat may cause laser refraction to occur (similar to the generation of a mirage), which can cause inaccuracies in laser distance readings. Under such conditions, line locations should be surveyed during the cooler hours of the morning.

Before commencing a survey, part of the line location planning exercise is to determine the degree of positioning accuracy required and if local benchmarks have that accuracy. If the benchmarks do not have the desired accuracy, then at least two benchmarks that do must be established in the area to be surveyed. The line drawn between these two benchmarks is called a *baseline*, with chosen points along the baseline being considered as newly established benchmarks of the required accuracy. Distances from these baseline points to stations on seismic lines may then be determined. A baseline provides the surveyor with secondary benchmarks from which the seismic lines may be referenced, and therefore more flexibility for surveying the seismic lines.

Alternatively, if three benchmarks are located in an area, their three connecting baselines provide three sets of secondary benchmark locations. Since three points form a triangle, this is sometimes referred to as constructing a *triangulation network* of benchmarks across the length and breadth of an area. For example, in Figure 139, point P on a seismic line is to be surveyed. It is located within a triangle of stations A, B, and C. Distances from P to the three

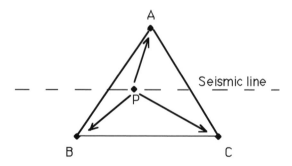

Fig. 139. Triangular network.

benchmark stations or to chosen points along the baselines can be measured, and an accurate location of point P therefore can be obtained. It is more common to use the three benchmark stations as the primary survey benchmarks than to use the secondary baselines.

When measuring distances from seismic lines to two benchmark stations (known as *tying in*), the accuracy is limited by the accuracy of the each of the two benchmarks. Tying in with a triangulation network improves the accuracy, with three sets of distance data for comparison and checking.

Vertical elevation values can be obtained from station to station by viewing a stadia rod from the DMU and computing the elevation value. Some DMUs compute elevation automatically when the appropriate information is supplied to them. The true value of elevation above sea level is not critical, but a relative value from station to station is needed for computation of elevation statics corrections. A 0.5 m elevation difference can cause a 1-ms arrival-time difference from one station to the next, which if unnoticed can cause a loss of subtle geologic features in the stack of the seismic data. Elevation accuracy can be checked with a loop closure map (Figure 140).

The map is produced by calculating and annotating elevation changes between all line intersections. Arrows may be placed between intersection points in the direction of decreasing elevation; sometimes a rotating arrow is used within each loop to indicate the direction of a total sum of elevation changes. If there are no elevation errors, summing the elevation changes around any closed loop gives zero. Nonzero sums are caused by measurement errors or computational errors. An unjustified change in any one branch

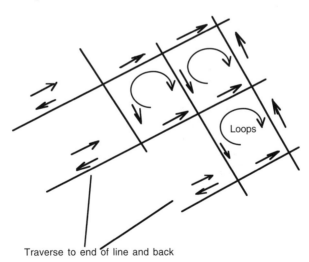

Traverse to end of line and back

Fig. 140. Loop closure map.

will cause a change in all associated loops; this can be used as a quality control scheme for the surveyor. On lines on which loops do not close, a *double run*—a traverse to the end of a line and back—is used. Double runs are shown at the left of Figure 140. A difference in total distance between the run to the end of the line and back will indicate an error in distance measurement that must be corrected.

5.5 Satellite Surveying

While land surveying has been performed for many years using theodolite, rod, and chain, new technologies in surveying have evolved in recent years. Navies around the world have used satellite systems to position their ships; in the early 1970s such systems became sufficiently accurate to use in the offshore seismic business. Magnavox Inc. was a leading commercial retailer of satellite receiver systems that allowed seismic ships to obtain their positional information independent of shore-based systems.

Magnavox receivers were able to use the TRANSIT satellite system, which was operated by the U.S. Navy. Basically, six satellites are in polar orbit 1075 km above the Earth, each one circling the globe every 107 minutes (Figure 141). The satellites form a cage within which the Earth rotates, carry-

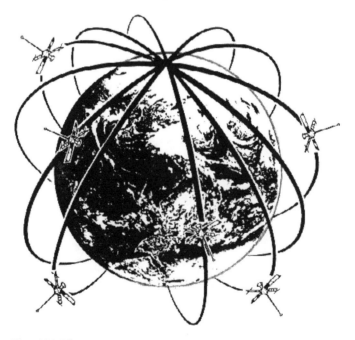

Fig. 141. The TRANSIT satellite system.

ing us past each orbit in turn. When a satellite passes overhead, a pass is said to occur during which a fixed position, also called a *fix*, is measured. The average time interval between each satellite pass is from 10 to 110 minutes, depending on the latitude of the receiver. Polar latitudes have the shortest intervals between fixes. Each satellite transmits two carrier signals, 150 MHz and 400 MHz. Every two minutes, each satellite transmits a message sufficient for a receiver to determine the receiver's location at that time.

To obtain the receiver's position relative to the satellite, the *Doppler shift*— a function of the observer's position and the satellite motion—is measured. A frequency shift is measured which is a result of satellite movement toward the receiver, i.e., more cycles per second received than transmitted in a given time. For each wavelength the satellite moves closer, one more cycle is received. As the satellite moves toward the ship, the received wavelength from the satellite appears to compress so that its apparent frequency increases in value (like an ambulance siren increasing in pitch as the ambulance travels toward an observer). When the satellite passes the ship, the received wavelength appears to stretch, causing the apparent frequency to reduce in value (like an ambulance siren pitch decreasing as an ambulance travels away). Thus, a *Doppler count* is a measurement of a change in satellite apparent frequency, and that apparent frequency depends on the receiver's location. By making many Doppler counts, the receiver's location can be computed accurately. About 20 counts, taking 10 to 18 minutes, provides a positioning accuracy of about 200 m.

All satellite receptions are subject to errors in transmission (such as those caused by atmospheric changes or equipment failure). To eliminate reception errors, dual receivers may be used in the *translocation mode*, in which one receiver is placed permanently on a benchmark (first order) while the other receiver is located at the position whose coordinates are desired. Any error found between the benchmark receiver's computed position and the true benchmark position can be used to correct the position computed by the other receiver.

TRANSIT receivers have been used to establish geophysical and navigation-station locations, but their low-order accuracy limits the positioning accuracy of any subsequent seismic survey. TRANSIT satellite receivers are used on seismic survey vessels, but because a single fix is accurate to within only 200 m, they are too inaccurate to provide total control of survey location. Consequently, when the U.S. military began to establish a far more accurate system in space, known as the Global Positioning System, the industry was quick to adopt it.

5.5.1 Global Positioning System (GPS)

The NAVSTAR (Navigation System with Time and Ranging) Global Positioning System (GPS) was conceived in 1973 when the U.S. armed services agreed to cooperate in the development of a highly accurate, space-based, military satellite navigation system orbiting just under 19 000 km above the Earth twice a day. The system was designed to provide instantaneous positioning for navigation purposes 24 hours a day.

The system consists of three segments: (1) a space segment comprised of satellites that transmit coded radio signals, (2) a ground-based control segment to monitor the position and health of the satellites, and (3) the user-equipment segment. When fully operational, the GPS system will have 18 satellites in orbit, with three spare satellites available for use if required. This will ensure that four to seven satellites will be visible from any point on Earth at all times.

Each GPS satellite transmits a unique signal on two L-band frequencies. The satellite signal consists of the L-band carrier wave modulated with a precision or *P code*, which can provide positioning accuracy to better than 20 m. The P code can be encrypted for military use only. A less precise *clear acquisition* or C/A code provides accuracy to better than 100 m. This code is available for public use. The primary function of the P and C/A codes is to permit the signal transit time from the satellite to the receiver to be determined. The navigation message contains information on the satellite's instantaneous position and on the other satellite positions to enable the receiver to compute the other satellite times of appearance over the receiver. The transit time of a satellite, when multiplied by the velocity of light, gives the range from the satellite to the receiver. By receiving up to four satellites, the user can solve four equations with four unknowns: the 3-D position components of the receiver and the receiver clock error.

The GPS system can be used in both the *point-position mode* and the translocation or *differential-operation mode*. Because of orbital and timing errors, the point-position mode has an accuracy of 5 to 10 m, which is not adequate for land survey purposes. Point-position mode is used mostly for offshore navigation purposes. As with the TRANSIT satellite system, much higher accuracies can be obtained from observations made simultaneously at two ground receivers. In the differential mode, a number of the dominant errors will cause nearly equal shifts in the estimated positions of both receivers so those errors can be removed. This can provide accuracies of 5 m or better, with a separation between the two receivers of up to 500 km. If the two receivers are stationary for 30 minutes, their positions can be determined with an accuracy of 5 mm plus 1 mm per kilometer of separation.

The GPS system is a vast improvement over the TRANSIT Doppler system because it gives faster and more accurate results. Presently, the receivers available are becoming less expensive and bulky than the first-generation receivers. One example of the miniaturization of the receiver is the Magellan NAV 1000, which can be held in the hand, as shown in Figure 142. In general, a single GPS satellite fix will give a position no better than 30 m. A minimum of three satellites are required to compute a 2-D position, and four are required for 3-D (which includes altitude) positioning in latitude/longitude or UTM. Once the first fix is computed, updates can occur approximately every three seconds.

The GPS system is commonly used offshore for 2-D and 3-D surveying and rig positioning. It is also used in land surveying where immediate results of surveys may be plotted out at the end of a day's work. Both Germany and the Commonwealth of Independent States also have navigation satellites orbiting in space. If the U.S. military decides to commence encoding transmissions from the GPS system, we can expect to use German and Russian satellites instead as our positioning and navigation system. An alternative to satellite navigation is the shore-based radio-controlled system, discussed next.

5.6 Radio Navigation

All radio navigation systems are based on a simple principle: The distance covered by a radio transmission between two points can be measured by multiplying the transmission speed (the speed of light—299 792.458 km/s) by the transmission traveltime. In practice, however, there are several complications. Because radio waves reflect and refract in the atmosphere, a transmission path may not be equivalent to the direct path (i.e., the straight line) between the transmission points (Figure 143).

Furthermore, changes in atmospheric conditions can change the transmission path from one hour to the next. Sometimes radio waves may follow several transmission paths between two points; the path with the shortest traveltime may not be the signal received most strongly. This multipathing effect is known as *sky-wave disturbance* or *tropospheric scatter* and can cause the navigation system to be unacceptably inaccurate.

The basic principle of radio navigation can be used in two fundamental ways. In one scheme, known as *phase comparison,* the radio receiver detects phase differences among synchronized continuous signals broadcast by two or more shore-based stations. In some phase-comparison systems, a transmitter at the receiver location transmits a continuous signal that is rebroadcast by the shore-based stations. The receiver then detects the phase differences between the transmitted signal and the rebroadcast signals. In the other

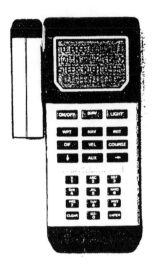

Fig. 142. Hand-held Magellan NAV 1000 GPS receiver.

scheme, known as *range measurement*, the ship's radio unit transmits a coded message to the shore-based stations, which then rebroadcast it. Because the signal in this system is not continuous, the radio receiver can measure directly the round-trip traveltimes—and, hence, the distances—from the ship to the shore stations. By contrast, in phase-comparison systems, only phase differentials are detected. That information has to be integrated (a process called *lane counting*) before it can be used to determine the ship's position.

If the distance to a single station is known, all that tells us is that the ship is somewhere on the circumference of a circle of known radius from the station. When the distances from a ship to two stations are known, the ship must be located at one of the places where the two circles associated with the distances and stations intersect (Figure 144). Usually, the ambiguity can be resolved by knowing the approximate ship position and picking the circle intersection closest to that position. Thus, with both radio navigation types, the location of

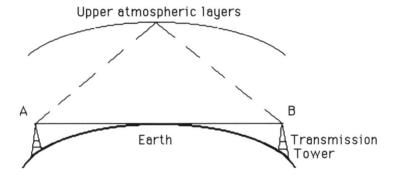

Fig. 143. Line-of-sight and tropospheric transmission.

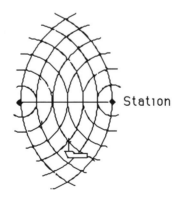

Fig. 144. Two-station fix positioning.

a ship is determined by finding the distance from the ship to at least two stations; this is known as a position fix. Additional stations provide redundant
data and, hence, greater accuracy of a fix position. In the situation where the
ship is near the line connecting the two stations, a third distance measurement is required to resolve the intersection ambiguity. Circular lines are used
to represent lines of equal distance to each station. These are referred to as
lines of position (LOPs). The minimum number of stations that can be used for
a fix is two (Figure 144), whereas a third station would allow a *three-way fix* on
a location.

5.6.1 Phase Comparison

There are two kinds of navigation based on phase comparison: hyperbolic
and circular. Of these, hyperbolic navigation is used most often. With the
hyperbolic network system, only the shore-based stations transmit a signal to
the mobile receiver. Continuous, uninterrupted analog signals are received
aboard the survey ship from two shore-based transmitters. The separate signals are synchronized so that the ship's mobile receiver can determine their
phase difference. Knowing the phase difference constrains the location of the
receiver to lie along a hyperbolic LOP (Figure 145). If a second pair of shore
stations are received and a further phase comparison made, a second LOP can
be obtained. The receiver is located where the two LOPs intersect. Normally,
only three shore stations are used unless the survey extends out of the range
of one station, in which case a fourth would be necessary.

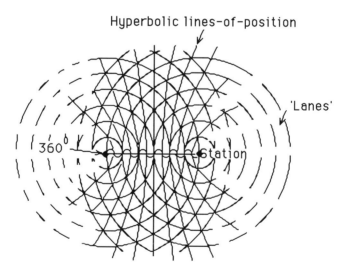

Fig. 145. Hyperbolic lines of position.

A measured phase difference is ambiguous by $+360n°$ or $-360n°$, where $n = 1,2,...$. Thus, determining the phase difference between two broadcast signals actually constrains the ship to be along one of many possible hyperbolic LOPs. This ambiguity is resolved by locking the phase of the receiver at a point of reference having a known, unambiguous phase value. Often, this is accomplished by steering the ship alongside a marker buoy located at a known phase.

Once receivers have been referenced, a ship may move away from the reference point and the receiver shows the phase changing according to the distance and direction traveled. When the phases pass through $360°$, a lane has been traversed. The number of times this happens is called the lane count. Typically, lanes are about 92 m wide. The phase difference (Figure 146) is referred to as the fractional lane count. The phase comparison receiver therefore is constantly monitoring the phase of the received signal and determining the LOP. If a signal weakens and the receiver loses lock on the phase, the receiver may skip a cycle, thereby adding or subtracting a whole lane. Thus, phase measuring requires continual position checks to keep track of the correct lane count.

When a seismic ship uses two shore stations, the ship is limited to the two sets of lanes. This form of operation has been typical of the North Sea surveys prior to the introduction of accurate satellite surveying. Typically, the stations would be located on the coast of one country (such as Britain) and the survey ship would then work off that coast. Alternatively, if a third station were added (which could be located on the Hook of Holland, for example), the survey ship would have LOPs from all three stations and improved navigational accuracy. If one station failed for some reason, the other two stations still

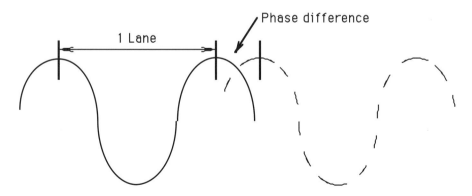

Fig. 146. Lane count phase difference.

could be used. Phase-comparison systems are useful for long-range work off-shore (up to 1000 km) and can offer an accuracy of better than 20 m.

Phase-comparison systems use tall shore-based transmission towers with electrical wires spread across the tower site, making the tower and ground the active radio transmitter. By comparison, range-measurement systems operate with smaller towers where only their antennae are the transmitters (so they may be operated offshore or on land).

Characteristics of the phase-comparison system are:

1) Unlimited number of users with hyperbolic networks.
2) Transmission distances of up to 400 km can be attained during day-time.
3) Long-range accuracy (theoretically 0.01 lane on the baseline).
4) The mobile unit must be calibrated at a known reference point to establish the lane count.
5) Shore stations are located at or near the shoreline to minimize trans-mission losses over land areas.
6) Transmitters operate in the medium frequency band of 1500 to 3000 kHz at long wavelengths, using whip antennae.
7) Circular networks can be obtained when an atomic clock standard is used as a frequency reference.

5.6.2 Range Measurement

These systems (known as *range-range* systems) use two or more shore-based stations (sometimes referred to as "beacons"). The ship-based transmitter/receiver unit transmits a coded message that is received by a shore-based station and then retransmitted to the ship's unit. The ship's mobile receiver computes the two-way traveltime and, therefore, distance or "range" to the shore station. At least three stations are needed to provide a position fix with the minimum accuracy required for seismic surveying.

The range-range equipment may allow the use of coded pulses so that up to 30 stations may be addressed at any one time, and as many as five mobile units may be used. Provided pulses are correctly synchronized, all mobile units may operate at the same time. This allows a number of ships to work in the same area simultaneously using the same base-station network.

The range-range system is capable of line-of-sight or just-over-the-horizon work and may provide a positioning accuracy of 1 to 30 m. However, range-measurement equipment operates on higher frequencies than the phase-comparison systems and rarely achieves distances of 300 km offshore. The range-range system is less expensive than the phase-comparison system because reference locations are unnecessary and larger towers are not a requirement. A further bonus of the range-measurement system is that the ship does not have to take time to sail back to a reference point as a phase-comparison ship must each time the phase has been lost.

Characteristics of range-measurement systems are:

1) Mobile units transmit signals to each shore beacon, which responds with a similar transmission back to the mobile unit. The distance is determined by the two-way traveltime. The range is usually limited to a line-of-sight distance.
2) Accuracy of 3 m at 10 km and 50 m at 200 km is possible.
3) Transmitters operate in the UHF/SHF frequency bands at 400–6000 MHz at short wavelengths using Yagi-type, high-gain directional antennae.
4) High-gain directional antennae can be employed for improved performance or a reduction in power.
5) Pulse correlation methods can provide improved signal-to-noise characteristics.
6) Circular lines of position provide simple position computations.

When a range from a ship to each of two stations is known, it is possible to calculate the receiver location, which will be either side of the direct line between the two stations (known as the baseline). In Figure 147, two island-based stations, A and B, are transmitting to a ship. When only two ranges are measured, the receiver position is constrained to be in two possible locations, corresponding to the two points at which the circular lines of position intersect (Figure 147). Addition of a third range to station C resolves this ambiguity (Figure 148).

For the best accuracy of positioning using three stations, the survey program should lie within an area that subtends an angle of 30° to 150° to every

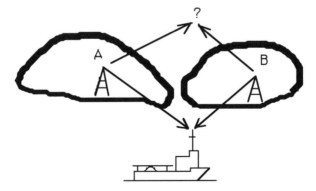

Fig. 147. A two-way fix of a position.

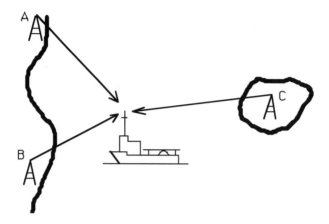

Fig. 148. A three-way fix of a position.

pair of the three stations (often referred to as an area of *good angle*), as shown in Figure 149.

To construct a good angle area from two stations, put a compass point on station 1 (Figure 149) and draw an arc from station 2 to both sides of their baseline. Now put the point on station 2 and draw the arcs from station 1 either side of the baseline to intersect at locations A and B. With the point on intersection B, draw the right-hand part of a circle from station 1 to station 2. The lines drawn from any point on this circle to stations 1 and 2 subtend an angle of 30° (known as the *Euclidian rule*). Now put the compass point on the other intersection point (on the other side of the baseline) at A and draw an arc from station 1 to station 2. The lines drawn from any point on this arc to stations 1 and 2 subtend 150°. The shaded area between the 30° intersection

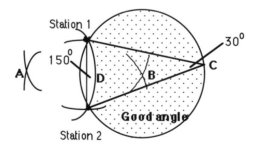

Bad angle of cut

Fig. 149. Angle-of-cut construction for range-measurement systems.

and the 150° intersection is called good angle because the ship positioning has the required accuracy for offshore seismic survey work. Any position outside of this area (unshaded area outside of the circle) is considered *bad angle* and of too low an accuracy level of positioning for seismic operations.

With the line-of-sight range systems (i.e., within 100 km), the range maximum is estimated from

$$R = (K\sqrt{h_1} + K\sqrt{h_2}) \times 10^3, \tag{37}$$

where R = range in km, h_1 = antenna height above sea level in m, h_2 = mobile antenna height above sea level in m, and K = an empirical factor in units of meters$^{1/2}$. For example, if $h_1 = 9$ m, $h_2 = 100$ m, then $R = 13$ km with $K = 1$. K varies with different navigation equipment from 1 to 2.5, and depends upon the transmitter power, the operational frequency, and the atmospheric propagation conditions in which the equipment is being used. Being empirical, such factors are only approximate and are generally determined by the operator after tests in an area of operation.

5.6.3 Other Radio Navigation Systems

Although GPS is very popular, a number of other navigation systems are often in use. A brief description of such systems follows, with briefer notes on some of the systems popular in the past but now archaic.

Syledis—Manufactured by Sercel of Nantes, France, Syledis operates in either hyperbolic or range-range mode with virtually unlimited mobile users. It has a narrow (1 MHz) operating bandwidth and a good low-noise signal. Operating frequencies range from 420 MHz to 450 MHz. The range of Syledis is approximately 130 km without booster amplifiers and up to 250 km with power boosters and directional antennae. Accuracy and repeatability values of a few meters are possible at the maximum range. When Syledis is used with linear amplifier boosters to increase its maximum range, there may be a loss of accuracy. Base stations do not have to be manned, and the system is very portable. Syledis is one of the best medium-range radio-positioning systems available.

Maxiran—This is a range-range system using a mobile unit designed to operate with three base stations. The system is multiuser and can operate with up to six mobile stations at one time. The maximum operating range is dependent upon the equipment configuration used and the prevailing atmospheric conditions. Ranges of up to 340 km are possible with high-gain directional antennae and linear amplifier boosters. The operating bandwidth of 6 MHz can make this system difficult to license near populated areas.

Mini-Ranger—This is a time-interrogation system developed by Motorola for short-range, high-accuracy applications. The operating range of the system is limited to line-of-sight distances because of the very high operating frequency of 5.4 GHz. The mobile unit transmits at 100 W and the shore-based beacons at 13 W. Dual-channel reception permits continuous display of distances to two base stations.

Trisponder—The Trisponder, built by Del Norte Technology of Euless, Texas, is similar to the Mini-Ranger. The principle differences are that it has higher operating frequency, better resolution, and 10 W mobile transmitter power.

Argo—Argo is a phase-comparison system. The name was derived from Automatic Ranging Grid Overlay system and was first introduced by Cubic Western Data Corporation of San Diego in 1971. Argo operates with up to four base stations in either the circular or hyperbolic modes, or a combination of both. Typical operational ranges are 750 km during daylight hours and 400 km at night. Accuracy is quoted as 0.05 lanes with lane widths between 75 and 94 m. Frequency range is from 1.6 to 2.0 MHz. As with all medium-frequency systems, sky-wave interference is a problem.

Decca Hi-Fix—Decca Hi-Fix is a phase-comparison system derived from the fact that a series of up to six stations (known as a "chain") can use the same single-frequency signal. Each station sends the transmission signal for a short period of time according to the timing sequence controlled by the master station. The sequential signal train has a repetition rate of 1 Hz. The geometric configuration can be either hyperbolic or circular or in combination. The lane width varies from 30 m at the highest frequency to 94 m at the lowest frequency.

Decca Mainchain—This is the original hyperbolic system developed in the early 1940s as a phase-comparison system with three base stations (one master and two slaves).

Pulse/8—This system was developed by Decca. The pulse repetition rate varies from 20 to 100 ms for identification purposes between chains. Pulse/8 can be used as a hyperbolic or circular system. The system derived its name from the eight pulse codes that are transmitted.

Loran—The name was the acronym for Long Range Navigation, a system which determined location by measuring the difference in arrival of pulse signals from a number of fixed transmitters. Loran "C" operated at a frequency of 100 kHz with a ground-wave coverage of 1600 km over land and 2500 km over water. The base stations operated synchronously, and the network grid was hyperbolic. An atomic clock was used on the mobile unit for circular network operations.

Nano-Nav was a phase-comparison system operating in the 1.5 to 2 MHz frequency band. Accuracy/repeatability was quoted as 0.01 lane width with lanes of 180 m on the baseline. Maximum range was 650 km during daylight hours and 450 km at night.

Raydist—This was a phase-comparison system that provided a circular network using two base stations, and it used a cesium frequency standard. The mobile unit contained a transmitter that sent the synchronizing signal to the base stations.

XR-Shoran—The extended-range Shoran system was a range-range system using three base stations, limited to ranges of less than 150 statute miles. XR-Shoran operated in the 230 MHz to 450 MHz range with an accuracy and repeatability of approximately 25 m under good conditions. The solid state version of this became Maxiran.

5.7 Navigation Systems

During marine seismic survey operations, the recording and processing of navigation data is handled by a computer. The numerous sensors monitoring the vessel's location and trailing equipment, together with the computer, form the navigation system. For example, if GPS is the primary operational navigation system, the secondary system may be Syledis supplying data to the computer, with peripheral equipment such as streamer compasses, acoustic equipment, the ship's gyrocompass, attitude sensors, and Doppler sonar providing other positioning data. The navigation system integrates all of this information to produce real-time estimates of the ship's position as well as the source and receiver positions.

Having these real-time estimates, the navigation system computer controls the ship's speed, its direction, and the shot firing time so that the actual seismic positioning data match, as well as possible, the intended, or "preplotted," acquisition program. Computer-generated preplots can contain a variety of useful information, either in graphical form or in tabular form. Usually preplots show at least the beginning and ending shotpoint coordinates for each planned line in an acquisition program. Sometimes every shotpoint position is plotted or listed. If shore-based radio stations are tabulated, the preplot often lists the intended ranges between each station and each planned shot position (in Table 5.1, for example, ranges are shown for stations A, B, C, and D). That information lets the navigation system operator know when a station will become unusable because of excessive range from the station to the seismic vessel.

Before the advent of computers, navigation of a vessel toward the seismic line was performed using preplots alone, which resulted in inefficiencies and lost time steering a vessel into position to begin a line. Today, steering is based on information provided by computer monitor screens that visually indicate the vessel's actual position and desired position.

Table 5.1. Seismic line preplot.

SYSTEM NAME			SYLEDIS STATIONS			
PT NO.	LATITUDE NORTHING	LONGITUDE EASTING	A (km)	B (km)	C (km)	D (km)
17	2 07 48.58 N 665239	018 24 34.74 E 533863	41.744	36.224	55.740	68.643
25	2 07 53.37 N 665387	018 24 40.61 E 533863	41.562	36.033	55.619	68.561
33	2 07 58.16 N 665536	018 24 46 48 E 534131	41.379	35.842	55.498	68.480

Below is a summary of calculations performed by the navigation computer:

1) Determine vessel's x,y coordinates or latitude and longitude. Bearing and distance to an aiming point such as the beginning of the seismic line, or the next shotpoint
2) Distance and velocity of vessel and seismic cable along course
3) Distance and velocity of vessel and seismic cable off course
4) Shotpoint number and firing control sequencing
5) Ship's track, cable, and source tracks
6) Positions of sources, receivers, and common midpoints
7) Position fixes and error probability
8) Coordinate conversions (x,y to latitude, longitude)
9) Datum conversions
10) Calculation of preplots onboard
11) Amount of coverage or fold along line, and CMP bin location points in 3-D recording

Integrated navigation systems have evolved continuously as computer hardware has advanced and new navigation system devices have been developed. When the TRANSIT satellite system first became available for commercial use, the geophysical industry rushed to it as a panacea for worldwide, 24-hour-a-day tracking with 50-m accuracy. It was learned quickly that long time intervals between satellite passes and a lack of accurate knowledge of the ship's velocity made the satellite system by itself almost useless for seismic navigation. Satellite navigation became useful only after the addition of many additional sensors on board the vessel and a computer to combine all available data into an integrated system.

When satellite positioning (TRANSIT, not GPS) is the only navigation system and shore-based stations are not available, a gap between satellite fixes requires the additional sensors to provide data to help compute the vessel's location (known as *dead reckoning*). Parameters that help to determine a ship's dead-reckoned position include the vessel's pitch, roll, yaw, and relative velocity between fixes. An accurate bearing (track) and velocity of the ship is thus required between each fix. To achieve this objective, the peripheral sensors installed on the vessel (the satellite receiver, sonar, inclinometer, velocimeter, gyrocompass, shore-based radio systems, rubidium frequency standard, and inertial guidance system) provide the data, which are integrated by the navigation computer system.

A ship's velocity is continually computed by the navigation computer by averaging the ship's position data over time. Alternatively, a sonar transceiver may be positioned on the ship's hull, providing a measure of the ship's velocity in two dimensions relative to the water bottom. The sonar system utilizes the principle of a Doppler frequency shift where the returned sonar signal is shifted up or down in frequency depending on the ship's motion. An inclinometer is used to sense the ship's deviation from a level position to correct the sonar measurement. Sonar data are notoriously inaccurate when the sea swell causes turbulence (as a result of aeration around the transceiver head). Sonar bottom tracking fails in water depths greater than 200 m. Because of these problems, sonar is used only as backup data to the primary systems. The velocity of sound propagation in water varies as a function of temperature and salinity, and so a velocimeter is usually incorporated in the integrated navigation system.

Streamer compass sensors provide orientation of the streamers with respect to some reference azimuth. Transponders mounted on the streamer, as well as on guns and tail buoys, provide further positioning information to locate the sources and receivers. In the case of 3-D surveys, tail buoys may even have GPS receivers mounted on their reflective mast. The ship-based gyrocompass provides the azimuth information for heading and navigation calculations. Gyro sensor data are also used to refine data provided by the acoustic systems used to locate the sources and streamers. Since the gyro sensors (like the sonar) are mounted on the hull, corrections for hull orientation become important.

Numerous errors can exist in the sensors used to measure the ship's velocity. The performance of an integrated satellite navigation system can be improved by coupling in shore-based radio positioning data. Navigation data are supplied continuously even if one or all sensors are out of order. The computers use a priority order to select sensor data. An abstract representation of the ship's position, attitude, and motion is maintained within the navigation computer. The data from each sensor type update that model in a way that automatically adjusts for the importance of the data relative to that from other sensors.

Integration of all data provides much redundancy. Thus, a typical operation can afford the luxury of a few malfunctioning sensors, and perhaps even a temporary loss of an entire subsystem. This integration is also very important for quality control. By comparing redundant data, the navigation system can pinpoint sensors that are misbehaving and automatically exclude them from the position calculations.

5.8   Navigation Planning for an Offshore Program Using Radio Positioning Systems

Planning an offshore program begins with a number of fundamental considerations:

1) The distance of the seismic program from shore. This will determine the required range for any radio navigation equipment.
2) The degree of positioning accuracy required
3) The survey time and duration (and hence the equipment availability)
4) Cost considerations
5) The number of vessels to be used (normally one)
6) The survey-area water depth

Let us examine the decision-making process with regard to each of these considerations:

5.8.1 Distance

If the majority of lines are short range (less than 100 km) from shore, then short range-range equipment may be considered. If the seismic program is greater than 100 km or less than 150 km from shore, medium range-range equipment with possible amplification may be necessary. If ranges are greater than 150 km offshore, longer range-range equipment using phase-comparison techniques may be necessary.

If offshore platforms are near the program area, a navigation system can be established by installing base stations on them. However, a consideration and constraint when using platforms is whether they provide good-angle navigation from three platforms for the survey area.

5.8.2 Accuracy

Accuracy depends on the precision and propagation errors (travel path and velocity) and the distance over which the signal is to be transmitted to the shore. Today the most favored range-range system for accuracy would be the Syledis system, which can provide submeter positioning accuracy on ranges up to 70 km. At longer ranges, booster amplifiers sometimes are required. These reduce the Syledis accuracy to less than 5 m.

5.8.3 Timing

Navigation contractors may not have equipment available immediately in an area of operation, and it may become necessary to hire it from overseas. If navigation equipment is not readily available, it may become necessary to conduct a survey using GPS alone, and tolerate any breaks in satellite reception. If base stations require satellite fixing or simple survey checking before use, a few weeks of notice may be required to mobilize and set up one or more GPS receivers.

5.8.4 Cost

Generally, the navigation cost is only 10–15% of the total survey cost, so system accuracy is the important criterion. If you have found two contractors offering the same system, you may discover one contractor may own his and offer it cheaper than the other (who hires it). If there is a choice of range-range systems of similar accuracy, then the relative age or usage of the equipment may become a factor. (If the equipment is old, a supply of electronic spares may be a problem.) In such cases, the higher cost of more recently manufactured equipment may be worthwhile. Also, when using a "term" rate (a day or monthly rental rate), vessel standby time as a result of navigation downtime may be chargeable to the exploration company (the cost of a seismic vessel shutdown may be $40 000 or more per 24-hour period). Therefore, having a standby mobile unit and at least one spare set of station transmitter/ receiver equipment (for example, a spare Syledis shore-based transponder unit) makes sense because standby rates for navigation equipment are cheaper than a seismic vessel's standby rate.

5.8.5 Number of Vessels

If two or more companies intend to conduct survey operations about the same time, it is cheaper to request a single contractor to perform the survey work as a total survey package for the two companies. Cost sharing of the same equipment can reduce mobilization/demobilization costs for separate survey crews. Most shore navigation systems allow the use of many vessels at the same time. Where there are vessels using the same navigation chain (for example, a supply boat may be moving a rig while a seismic survey is in progress), the cost of setting up the navigation chain can be shared.

Items that should be included in a shore-based positioning system's mobile and base units are listed below.

Mobile Unit

1 range indicator unit for each seismic vessel
1 spare indicator for each vessel
1 whip antenna for low-frequency systems
2 rotating beam antennae for each UHF unit
Batteries and/or gasoline power units, antennae masts as required
1 navigation computer with video monitors and magnetic-tape
    recording ability
1 line printer for each system
1 radio transceiver for communications with shore
1 complete set of spare parts for all of the above

Base Units

1 base station beacon for each shore location
1 spare beacon for every two locations
1 whip antenna for each low-frequency station
1 or more directional antennae for each UHF station
Batteries and/or gasoline power supplies
Antenna masts as required
1 or more vehicles as required
1 radio transceiver for communication with mobile
1 complete set of spare parts for all of the above
Local guards and helpers if needed

Exercise 5.1

Figure 150 is a land surveyor's loop closure map in which the elevation differences along portions of seismic lines are shown and errors in loop closure are shown in the center of each arrowed loop. Double-run elevation differences are also shown with a value for each run. Answer the following questions:

1) How many surveying errors greater than 1.0 m are shown on the loop closure map?
2) Are the loop–closure values shown consistent with the elevation changes?
3) Can any adjustments be made in the elevation changes to improve the loop closure values?

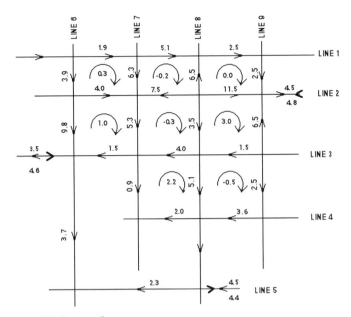

Fig. 150. Loop-closure map.

Exercise 5.2

(Show your map, station choice, and lines of cut.)

Plan a navigation network to cover the seismic program shown on the base map (Figure 151).

1) Select base stations to provide coverage over the prospect area, taking into consideration range capabilities of the equipment. Intersecting angles of the grid should be between 30° and 150°. A minimum of three stations placed at any chosen points on the coast is necessary.

2) Select the navigation system(s) to be used, indicating the total amount of equipment, personnel, and their cost structure for a survey to last one month.

Note:

We are fortunate that another survey has been run on an adjacent permit, using Argo and Syledis. Thus, Argo demobilization costs have to be considered (using three stations), and Syledis has been quoted as having no mobilization/demobilization costs. Consider the prospect is in the western Pacific, so for Maxiran use there *is* a mobilization/demobilization cost because that system is currently stored in Singapore.

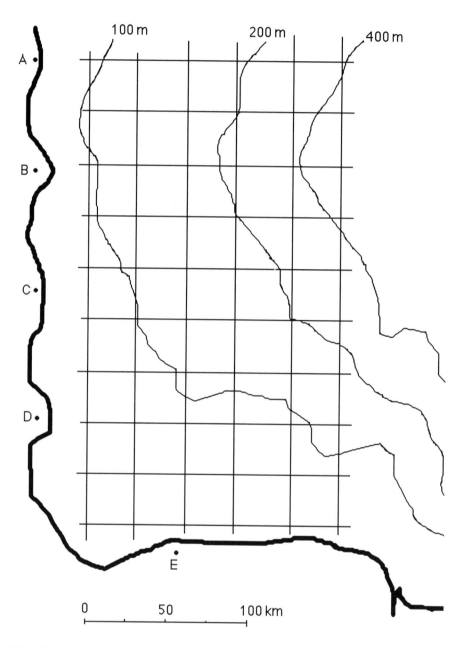

Fig. 151. Seismic program displayed on base map.

Maxiran maximum range is 250 km. Accuracy less than 20 m. Argo maximum range is 1000 km. Accuracy less than 25 m. Syledis without booster amplifiers range maximum at 130 km, accuracy approximately 5 m. Syledis with booster amplifiers range maximum at 180 km, accuracy approximately 15 m.

GPS satellite navigation may be used at any time, but its accuracy is questionable when there are few passes. Where signal is lost, however, the Doppler sonar bottom track may be used to a maximum depth of 350 m. The area of greatest interest for surveying and drilling is within the shore to a 300-m contour. There is no cost for the use of sonar.

3) Having established the best navigation chain for your survey:
   a) Does your selection remain best if the survey has had weather problems and takes one week longer than expected?
   b) Suppose that after the first 30 days equipment cost rates drop by 30%. Is the selection you made still the best, if the survey continues for an extra week?

Equipment Available	Rental/Day
Maxiran	
Indicator with 20 kW amplifier, navigation computer, track plotter, and antenna (mobile)	$400
Base station unit, 20 kW amp, power supplies, and antenna	$100
Mobilization/demobilization ex. Singapore	@ $400
Argo	
Three-range indicator, track plotters, and antenna	$380
Base station unit, power supplies, and antenna	$150
Demobilization only (since previous client paid the mobilization)	@ $2000
Syledis	
Three-range mobile, complete with power supplies, navigation computer, and antenna	$350
Beacon with battery, charger, and antenna (directional)	$80
Optional booster amp, 1 per beacon (more expensive than beacon since beacon is owned, whereas boosters must be hired for a minimum of 30-day period)	$100
Personnel	
Mobile operation (minimum of two required)	$200
Shore based engineer (minimum of one required)	$180

Chapter 6

Establishing Field Parameters

Before a survey begins in a new area, there should be a review of the local geology, past seismic surveys and well-log data, and seismic interpretation objectives. A knowledge of such data allows forward planning of the geophysical operations to ensure objectives are met. The local geology will provide information on geologic parameters that will affect the survey, such as the target depth, geologic features of interest, the maximum dip to be expected, and the level of resolution needed to image the target adequately. Past seismic survey data can act as a guide for the success of proposed geophysical parameters. For example, if one proposes to use a source at least as strong as that used in earlier work, and the earlier source successfully illuminated the target, then the proposed source also should succeed. Past survey data also can show where problems existed and how they were overcome and if economical data-quality improvements are feasible. Once the geophysical parameters have been established for a survey, they must not be changed without contractor approval from the client company. This chapter discusses how to determine the field parameters needed to meet the goal of recording the highest quality 2-D field data. The special requirements of 3-D surveys are considered in Chapter 7.

6.1 Survey Planning

Chapter 1 provided a general discussion of survey planning. In that discussion, target depth, seismic resolution, budget, and timing were featured because these are the areas of most interest to the field geophysicist. Once the target depth and features have been identified, the planner should decide what spatial and temporal resolutions are needed to illuminate the targets. The ability to image and resolve subtle geological structures puts a fundamental requirement on a minimum signal-to-noise ratio and the seismic signal bandwidth, which must be addressed prior to the field operation.

Thereafter, surveys must be performed within the allowed budget and within a time frame acceptable to both client and contractor.

The resolution requirements determine the maximum allowed line spacing and shot spacing and the necessary recording bandwidth. The planner should select a source that provides sufficient energy throughout the recording bandwidth.

Orientation of lines and line spacing depend on the structure size to be interpreted. Targets must be crossed at least once (ideally with at least one line over the structure in the dip direction). If the strike and dip directions are unknown, a uniform grid is used. For such a grid, an estimate may be made of the total survey size. One simple approach to determine the size of a survey is to make an estimate based on the area to be mapped, such as

$$\text{Total line km} = \text{Area} \left(\frac{1}{\text{dip line separation}} + \frac{1}{\text{strike line separation}} \right). \quad (38)$$

One approach to establishing the line separation in 2-D surveys is to determine initially the minimum size reservoir a petroleum engineer considers economical to produce. This will establish the areal extent of a field. Consider the producing field as circular so that a minimum-sized target of known radius is determined. At least three lines are required to delineate the two sides and the center of the target so the line spacing must be no more than the target radius, while tie-lines can be two times this line spacing.

If the targets are elongated (such as fault-bound sand traps), then the lines of most use for interpretation purposes are those perpendicular to the direction of maximum target length. These are referred to as the *dip lines* because they will run parallel to the geology of maximum dip. In this case, the maximum line separation would be half of the target length in the longest (elongated) direction.

Higher temporal resolution improves bed definition and detection of low-relief structural and stratigraphic changes. Because this may require a more costly source effort (for example, longer vibrator sweeps at higher frequencies), the cost-to-benefit ratio must be examined. High resolution may not be possible where high signal attenuation and absorption occur. Areas of steep geologic dip require higher spatial resolution and, therefore, closer receiver station spacing.

6.2 Noise Analysis

Interpretation accuracy is often limited by the seismic signal-to-noise ratio and signal bandwidth. At the start of a survey, a noise spread should be performed to measure the level of noise (be it surface wave, general geologic background, or sideswipe noise). If the noise is severe enough to hinder survey objectives, the geophysicist must decide how to handle the noise problem. One approach is to ignore noise problems in the field and to assume that various data processing steps (such as CMP stacking and the stack array explained in Chapter 2, and *f-k* filtering explained later in this chapter) subsequently will remove the problem of high noise levels. If the dynamic range between the amplitude of the noise and that of the underlying seismic signal exceeds the dynamic range of the recording instruments, the signal will not be recorded and the data processing methods have nothing to recover. Furthermore, prestack processing methods designed to operate on signal may be hindered by the noise. In such situations, it is best to select survey parameters to attenuate noise prior to recording in the field, but only if that can be done without compromising other aspects of the survey.

The most common methods of reducing noise in the field are frequency filtering in the recording instruments and wavelength filtering through use of directional source and receiver arrays. The wavelength data needed to optimize array parameters may be measured by performing *noise-spread tests.* These tests are described below. For each kind of spread, the offset between geophones and the source should extend from zero to the maximum offset that will be used in production recording. The receiver interval used during a noise test must be short enough to avoid *spatial aliasing* of the short-wavelength noise. Conventionally, it is set at 5 to 10 m; that is, all phones are bunched at 5-to-10-m intervals.

6.2.1 Spread Types

There are four methods of noise analysis: the *normal spread*, the *transposed spread*, the *double-ended spread*, and the *expanded spread*. All of the methods, except for the expanded spread method are used commonly by the industry prior to conventional land CMP recording. The expanded spread is used most frequently offshore by research agencies wishing to record deep crustal data using two vessels.

Normal spread—After laying out the geophone spread, a shot is fired from one end and the data recorded. The receiver spread is then picked up and moved to a new location one spread length away. Another shot is then fired from the same location as the first shot, after which the receiver spread is picked up again and the whole recording process repeated. The individual shot records are placed side by side, and continuous noise trains and reflections may be analyzed across the combined shot record. On the combined record, reflection events occur after refraction arrivals, followed by the ground-roll noise. This method is not popular because of the time required to move the receiver spread. However, it is ideally the best method for noise analysis, especially if the number of receiver recording channels is limited.

Transposed spread—The spread remains fixed in one location and the shot moves away from the receiver spread one spread length after each shot is fired. This method is more popular than the normal spread because it is easier to move the source than the receivers. A problem with this method is that a shot static difference misaligns noise and reflection traces when the individual shot records are placed side by side. However, the transposed noise spread is still the most popular type of noise analysis. It is often called a *walkaway* noise test because the shot or vibrator literally walks away from the receiver spread during the recording operation.

3) **Double-ended spread**—The transposed spread is repeated, but in this case, the shot is fired from each end of the spread, providing twice as much information and data than does a single-ended noise spread. It can be used to examine how noise would appear on split-spread records, or to examine the weathering effect on refracted arrivals, and it gives an early indication of the presence of dipping horizons.

4) **Expanded spread**—The expanded spread is often performed by using two vessels shooting in the opposite direction from each other. They can either move away from a starting point or they can move in opposite directions toward, then past, and away from each other. For example, in Figure 152, a vessel at shot "a" may be moving toward another vessel which is towing a streamer at spread "A" toward the first vessel. The next shot is fired a spread length closer, with the spread at "B," thereby duplicating CMPs while measuring surface noise from different offsets. Shots are moved along, continuing to fire, while the receiver spread is moved continuously in the opposite direction. Since the vessels can move long distances apart, the

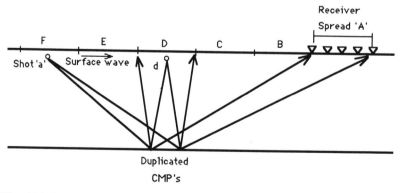

Fig. 152. Expanded noise test.

spread is called *expanded*. Such a recording method allows a signal-and-noise analysis, as well as limited CMP coverage over a length of the Earth's crust.

Noise created by shots can be classified as in-line or broadside. While most in-line noise is a result of ground roll within the weathering layer, broadside wavetrains can occur when source energy reflects into the receivers from subsurface features that lie outside of the vertical plane defined by the line direction. Such features include shallow carbonate or silcrete pillows within the weathering layer and fault planes that run parallel to the line. If such broadside wavetrains (commonly referred to as *sideswipe*) are suspected, a crossspread noise test in a "T" or "L" shape can confirm their presence.

If a wavetrain is a reflection from a steeply dipping event in the profile of the noise spread, it is desirable to retain it. If it is sideswipe, then it is desirable to attenuate it for 2-D surveys but retain it for 3-D surveys. When a "T" or "L" spread is recorded, sideswipe may appear as conventional linear noise on the receivers which are in-line with it. If the wavetrain appears to have curved moveout on the main in-line spread but is flat on the other spread arm, it may be a deep, steeply dipping reflection and should not be attenuated.

In years past, 2-D areal geophone arrays were used to attenuate broadside noise in the field. The linear array theory that is described in Chapter 2 applies to 2-D arrays, except that the apparent noise wavelengths to be attenuated arrive from different directions. In an attempt to attenuate noise from directions other than those along the line of receivers, experiments have been conducted using arrays spread over an area (known as *areal arrays*). Geophones may be laid along the seismic line direction and off-the-line direction. The *star array* and the *herringbone array*, as shown in Figure 153, are two common areal arrays.

The use of such arrays is limited, not only because of the increase in effort and cost required to lay geophones in an areal array but also because, once a particular lateral reflection event has been physically passed, it no longer presents a problem. Today broadside events that pass diagonally across the seismic record may be attenuated in the processing center using an *f-k* filter.

The noise-spread record indicates the energy level of received reflected signals and noise interference. The apparent wavelength of noise is used to determine the array's geophone separation in an attempt to attenuate the noise from whichever direction the noise may arrive. The theory for this form of attenuation was discussed in Chapter 2. Since a shot record may show many noise trains of different wavelengths, one solution is to use weighted arrays, which also are discussed in Chapter 2. However, areal receiver arrays have built-in weighting because of their 2-D layout, since the geophones are at unequal separations when viewed at different azimuths. Consequently, designing an areal array is more complex than designing a linear array.

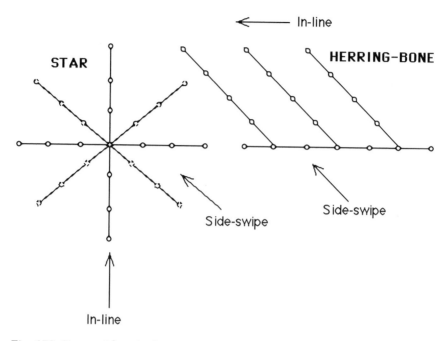

Fig. 153. Star and herringbone areal arrays.

Once the station spacing and receiver array configuration are determined, noise spreads may be repeated to determine the optimum source configuration, which may also be either linear or weighted.

6.3 Experiments for Designing Parameters

Noise spreads allow an estimation to be made of the optimum source and receiver array parameters, the relative signal strength of reflections, any variations in signal and noise frequencies with a change in offset, and the coherency of events. As discussed in previous chapters, arrays help attenuate coherent as well as random noise. Whether a particular noise appears as random or coherent to the elements of an array depends on the spatial coherence length of the noise and the element spacing. If array elements are spaced apart by an amount that exceeds the spatial coherence length of the noise, then summing N of those elements reduces the noise by $1/\sqrt{N}$. However, if the element spacing is less than the spatial coherence length of the noise, then summing attenuates the noise by a lesser amount. Noise-spread tests with closely spaced receivers provide a way to determine the spatial coherence of the noise. Such data can help one judge the benefits to be gained by adjusting the spacing between elements in an array. With the use of field computers, the performance of specific arrays may be examined in the field, and the arrays modified to improve their performance.

6.3.1 Array Performance Analysis Using 2-D Frequency Transforms

Figure 154 is a schematic plot of frequency f versus wavenumber k, displaying where signal-and-noise wavetrains typically may lie in the f-k domain. The f-k plot is a 2-D Fourier transform of a seismic record, plotting data in a frequency versus wavenumber domain. Beneath the f-k plot is an array response plot showing amplitude response versus wavenumber k (as used in Chapter 2). The k axis is the same for both f-k and array-response plots.

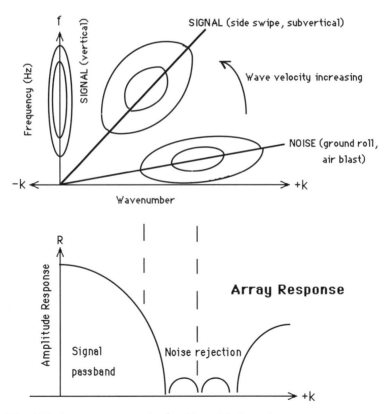

Fig. 154. Array response in the f-k and k domains.

The separation of the signal-and-noise wavetrains on the f-k plot is a result of a difference in apparent velocity. The schematic contoured regions of the f-k plot indicate that the signal strength is greater at the center of the contours. By arranging the two plots one above the other, the relationships of the two domains may be compared for the same k number values. For example, the array response plot shows that, with the correct receiver-array choice, noise trains may be rejected while most of the signal may be retained. This form of analysis is useful in determining array design in the field. When appropriate, computing facilities are not available, an approximate f-k plot can be performed by hand, just as can an array response. The following section will explain this approach in detail.

6.3.2 Crude f-k Plotting

A seismic field array can be considered equivalent to an f-k filter. This can be shown by a simple exercise with a synthetic field record. Suppose that the geophone spacing that will attenuate the ground-roll noise train in Figure 155 is to be determined. The synthetic field record shows a shallow reflection and the unwanted ground roll noise train arriving later with a lower velocity. The f-k plot for the record is sketched, and a simple two-geophone array may be designed to attenuate the ground roll as follows:

1) The center frequency of each arrival's wavelet is computed from the wavetrain's periodic time, as is normal practice (see Section 2.3.1). In this example, the center frequency for the reflection is $f = 1/t = 1/0.033 = 30$ Hz. For the noise train, $f = 1/0.07 = 14$ Hz.

2) Using each arrival's moveout, each arrival's velocity V and wavenumber k is determined $(k = f/V)$.
   For the reflection, V = distance/time = 100 m/0.01 s =10,000 m/s.
   Hence, $k = 30/10,000 = 0.003$ m^{-1}.
   For the noise train, V = 100/0.035 = 2857 m/s. Hence, k = 14/2,857 =0.005 m^{-1}. This is the value at which maximum attenuation will be desired.

3) The two events' values are plotted on the f-k plot, with a rule-of-thumb range of frequency being 0.5 and 1.5 times the center frequency computed from step 1 above (i.e., a wavelet with a center frequency of 60 Hz has an approximate range of 30–90 Hz). A line is drawn through the origin representing each wave, and contours are drawn spanning the approximate frequency range.

4) The spacing for two geophones now may be determined so that a notch in the f-k plot attenuates the noise while the array passes the signal. Consider that a maximum signal passband has a 100% response,

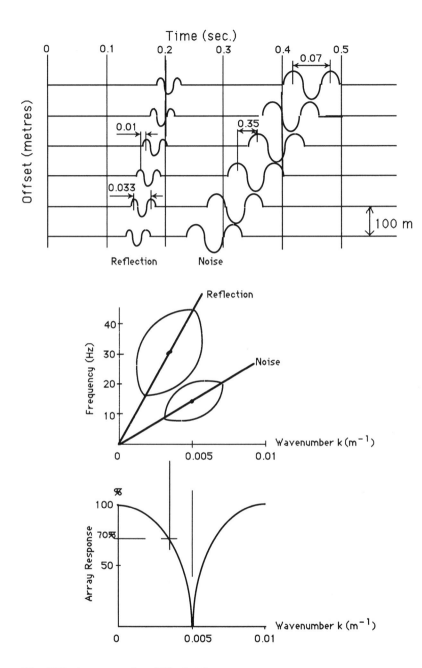

Fig. 155. An example of *f-k* plotting.

whereas the maximum attenuation is at a response of 0%. With two phones, a single notch occurs at $k = 1/nd$, i.e., $1/2d$. In this case, the desired distance $d = 1/2k = 1(2 \times 0.005) = 100$ m.

5) From the array-response curve, it can be seen that this simple array will pass 70% of the reflection at the center frequency while attenuating most of the noise. This provides a good reason to make the response curve passband steeper by the use of more phones with appropriate spacing and/or weighting.

6.4   Dip Recording and Beaming Effect

Most reflection events are not traveling vertically when they reach a geophone array. An array attenuates random and any coherent horizontal noise that coincides with the linear axis of the array, as well as the horizontal components of reflections which do not arrive vertically and which coincide with the linear axis of the array. The degree of attenuation is dependent on a reflection's apparent wavelength and the array dimensions (see Section 2.5). Apparent wavelength depends on the depth and dip of a reflector, the source-receiver offset, and the velocity structure between the reflector and the surface (the last because of ray bending at each reflecting layer).

Based on the theory of reciprocity (i.e., the interchangeability of receiver and source arrays), source arrays can be tuned to beam their energy in a preferred direction, just as geophone arrays can enhance data received from a preferred direction. This is useful when we wish to concentrate energy toward a direction that will best illuminate a desired target. For example, source arrays wide in the cross-line direction have been used in an attempt to reduce out-of-the-plane reflections (i.e., sideswipe), whereas in marine surveying, the surface ghost provides a significant beaming effect in the vertical direction.

Arnold (1977) examined energy source beaming with vibrator arrays, and in 1982, the author of this book compared the beaming effect of explosive cord with that of dynamite shot holes drilled in an array. Arnold found that energy beaming was partially successful but was limited by the complexity of the geology being profiled; the author of this book found that the longer the cord, the greater the reflected signal (irrespective of explosive weight per unit length), and that the cord was superior in beaming to the shot-hole pattern. If explosive cord is detonated from one end, the wavefront is tilted because of the finite detonation velocity of the explosive cord (about 7000 m/s). (With pattern-style shot holes, shots may be placed at varying depths to tilt the wavefront in a desired direction, or an electric detonator delay circuit can cause staggering of the individual shots in order to tilt the wavefront, as shown in Figure 80.)

Small charges also have been used for energy beaming. The technique is to load a single drill hole with a series of separate small charges, one on top of the other, separated by a meter of soil. An explosive cord taped to each charge is detonated by a single cap at the top of the hole; as the cord explodes (at a rate of 7000 m/s), each charge is ignited in turn. Theoretically, provided the spacing of the charges is correct, the downgoing energy should be increased as it travels past each exploding charge. Tests with small charges in this configuration have met with varying degrees of success; while there may be a near-field beaming effect, the concentration of energy in the near field does not travel to the far field, probably because of absorption and spherical spreading effects.

A short receiver array is preferred for shallow reflections because the reflected energy from those targets tends to have a large moveout over a short distance (a wide range of horizontal component of their wavenumber). To avoid attenuating this wide angle energy, a short array is needed. Energy from deep reflectors usually arrives almost vertically at the surface, displaying little moveout on shot records. A long array can attenuate noise without affecting the signal from deep reflectors.

Because of the effect of long arrays on signal from shallow horizons, the seismic industry has trended toward the use of short arrays in the field in recent years. This trend was made possible by the advent of wide dynamic-range recording systems capable of handling 200 or more channels. Sometimes shorter arrays are combined into longer arrays to reduce the volume of a prestack data set. The combination is done either by a field computer or in a data processing center. Usually, NMO timing corrections are applied prior to such trace summing based on nominal moveout velocities of the region. These corrections prevent the energy reflected from shallow horizons from being attenuated by the summation.

6.5 Parameter Optimizing

6.5.1 Receiver Frequency

Sometimes the amplitude of low-frequency ground roll is high enough to saturate the output of the geophones. When that happens, one can switch to geophones that are either less sensitive or that have a natural frequency cutoff (see Chapter 2) above the frequency of the ground roll. Either solution may compromise the ability of the geophones to detect low-amplitude, low-frequency reflection signal. Performing and analyzing noise tests with different geophone types can help a geophysicist to select the geophone parameters that are most appropriate in a given area.

Shallow, high-resolution seismic surveys can have a particular problem with high-amplitude, shot-generated noise because of the short shot-to-receiver offset distance. Instead of increasing the natural frequency of the geophones in an attempt to attenuate this problem, an alternative approach is to use a low-cut filter in the range of 200–300 Hz, with a 24 dB-per-octave roll off (Steeples, 1992). This assumes that the reflected energy is of sufficiently high amplitude and frequency content that reflections also are not attenuated.

6.5.2 Energy Source Parameters

In land and marine surveying, the most important source parameters to consider are bandwidth and power. The source must produce sufficient energy over a sufficient frequency range to allow reflections from the target horizons to be detected and resolved. Sometimes the noise producing or attenuating characteristics of a source are important. For example, if in a particular region only frequencies of less than 35 Hz penetrate to the target, then one may not want to use a source rich in higher frequencies because all that will do is contribute to shot-generated noise. Other source characteristics that can be important are wavelet shape, costs, reliability, environmental impact, and logistics.

The bandwidth and power of a marine air-gun source are governed by the number of guns in the array, the size of the guns, and their deployment depth. Typically, an air-gun array is built up of modular subarrays of six to eight guns. Since such subarrays are carefully designed to optimize power and bandwidth, it is usually best to select marine source parameters in terms of subarrays. For example, the strength of a three-subarray source can be doubled by adding three more subarrays (assuming the ship's air compressor capacity is not a limiting factor). As has been discussed in Chapter 3, the bandwidth of an air-gun source is controlled mainly by the depth of the air gun. A typical depth is 6 to 7 m, which puts the first nonzero frequency ghost notch at about 125 Hz. Shallower guns have more high frequencies in their signature but fewer low frequencies (because of the zero-frequency ghost notch). Deeper guns can be used to enhance lower frequencies while sacrificing the high ones. In general, the first nonzero frequency notch should be kept above the top end of the data processing bandwidth. Air-gun parameters other than depth and the number of subarrays fired are adjusted rarely once a basic subarray design has been chosen.

With land dynamite, however, there are many variables that require attention and site-specific optimization. For example, when using an explosive energy source, the frequency spectrum is related to the charge size. Generally, smaller charges produce sharper pulses and higher frequency content (see Figure 156).

A pattern of smaller charges produces much higher amplitudes and higher frequencies than does an equivalent single large charge. (This is true because of the one-third power law, just as in the case of air guns.) For example, in Figure 156, two of the small charges produce an amplitude equal to that of the one large charge but use only one-quarter the amount of explosive material.

Figure 157 shows six shot records using different charge sizes at three different offsets from a receiver line which had star, square, and in-line receiver arrays. The records have been displayed using the same fixed gain. One one-quarter-pound and eight one-quarter-pound charges of explosive were fired at an offset of 134 m (440 ft) from the receiver line. Both the star and square arrays canceled little of the coherent noise but provided a lineup of reflections which were weakest using the in-line array. This comparison showed that there was an increase in signal-and-noise energy as a result of increasing the charge size by eight times at the short offsets.

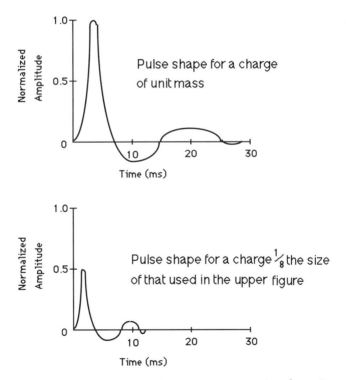

Fig. 156. Pulse duration and amplitude are proportional to the cube-root of the charge size.

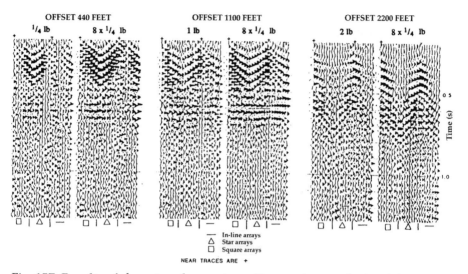

Fig. 157. Results of changing charge size, offset, and array design (after McDonald et al., 1981).

The source-to-receiver offset distance was then increased to 335 m (1100 ft) and a one-pound charge compared with the eight one-quarter-pound charges. The reflected signal strength was of greater amplitude and coherency, using the small charges (amounting to 0.9 kg or 2 lb), than the single charge (of 0.45 kg or 1 lb), as may be expected. At an offset of 671 m (2200 ft), a two-pound charge was compared with the eight one-quarter-pound charges (equal charge weight) and the small charges continued to be marginally superior to the single charge. The conclusion which may be drawn is that, in this case, a number of small charges are preferred at all offsets to a single larger charge, even though the signal-to-noise ratio may remain fairly constant.

When data have been recorded using an explosive source, a signal-to-noise ratio of 1:1 is acceptable for deep penetration, whereas 10:1 is preferred for high-resolution recording. This ratio can be found by comparing *fixed gain* amplitude playbacks of target reflections with noise records for which a shot was not fired. The noise records must be typical of the area and must not be recordings that were especially quiet.

6.5.2.1   Source and Receiver Depth

The depth at which a source is fired affects the frequency content of signal, noise, multiples (ghosts), and statics (if any). As mentioned in the previous section and Chapter 3, air-gun-array depth is extremely important since gun arrays must be positioned to minimize ghost interference with the recorded reflections. Marine streamers also should be placed at an appropriate depth to avoid surface ghost problems. An accepted mean for streamer operation has been 10 m, which is sufficiently shallow to keep ghost notches above the seismic bandwidth but sufficiently deep to avoid the increased noise suffered by shallow streamers. Streamers may be raised during recording operations to avoid a shallow seabed; this is acceptable since it moves the ghost notch to higher frequencies. However, the prevailing sea state then becomes an issue since sea turbulence can cause a rapid increase in streamer noise level. A long swell at a low sea state (say, 3) can cause more turbulence noise on streamers than a short swell at a medium sea state (of about 5). When sea state-induced noise exceeds noise specifications, a crew must shut down unless the client gives permission to continue. Streamers also may be lowered to avoid seaborne traffic. In both cases, the streamer noise level must be monitored in case it moves outside of noise specifications.

In land recording, receiver depth is not an issue, whereas source depth is only an issue with explosive sources. Ideally, charge holes should be beneath the LVL to minimize noise and high-frequency energy loss and to improve coupling. Ghost reflections, whether resulting from peg-leg reflections at the surface or within the weathering layer, tend not to be a problem on most land surveys. Since they occur at variable times (because of topography changes or variable shot depth), the CMP stacking process often cancels them. If ghosts are a problem, and if the LVL has a constant velocity, then the explosive depth may be adjusted. However, in practice, the LVL velocity is likely to vary along the line, so changing the source depth rarely resolves the problem of ghosts.

6.5.2.2 Recording Cable Geometry

When seismic energy arrives at reflecting layers at incident angles greater than the critical angle, it refracts rather than reflects (Figure 158). Consequently, if offsets are too large, the recorded seismic energy will be refractions rather than reflections. Although refraction data do have uses, in reflection

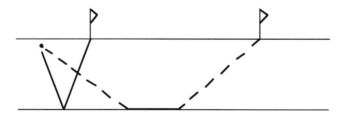

Fig. 158. The near offset must be short enough to record reflections rather than refractions for the shallowest events desired.

surveys, the far offsets should be small enough so that the target horizons are detected as reflections rather than as refractions. Conversely, if offsets are too short, the normal moveout correction may become too small for accurate velocity analysis and optimum multiple attenuation. Hence, it is important to determine both maximum and minimum near and far offsets for recording good-quality data.

The desired receiver cable length is a function of the depth of the horizon to be imaged. As a rule, the industry makes the far-offset receivers about the same distance from the source as the horizon is beneath the source. The near offset tends to be determined by the amount of noise that is generated by the source. A more rigorous determination of the desired near- and far-offset distances is based on the expression for NMO, as follows:

The far-offset distance is determined by both the critical angle and the amount of NMO stretch allowed on the reflections expected on the far-offset traces. NMO corrections applied to those traces will stretch the reflection events upward, causing a reduction in frequency content. Consequently, a decision has to be made regarding how much stretch (i.e., frequency reduction) is tolerable for an acceptable stack. The greatest stretch occurs on the far-offset traces of the shallowest reflections. Since the far offsets are most important for velocity analysis, then the maximum far-offset distance is mainly a function of the maximum NMO stretch on the shallowest reflections of interest. If the NMO correction δt stretches the data on the far trace an equal or greater time value than the period of that reflection's dominant frequency, then that stretched data will be unusable. This, therefore, sets the criterion for determining the far-offset distance. From the NMO equation (17) in Chapter 1, consider that the NMO correction

$$\delta t = t_x - t_1, \tag{39}$$

so

$$\delta t = \sqrt{\frac{x^2}{v^2} + t_1^2} - t_1, \tag{40}$$

where x = the far offset distance, t_x = two-way traveltime of the horizon containing the maximum stretch at the farthest offset, t_1 = two-way traveltime of the horizon containing the maximum stretch at zero offset, and v = the average velocity to that horizon of interest.

Stretching is quantified as

$$\frac{\delta f}{f} = \frac{\delta t}{t_1}, \tag{41}$$

where f is the dominant frequency, and δf is the change in frequency (Yilmaz, 1987). In estimating the dominant reflection frequency f, estimate on the low side. A stretch of the far-offset traces greater than 50% would distort the reflected data after stacking (i.e., if $\delta f / f$ is greater or equal to 0.5). Consequently such data are muted out before the stacking process. This maximum stretch allows us to set the maximum far offset.

For example, if the maximum NMO correction δt is 50% of a deep reflection's zero-offset two-way traveltime, then $\delta t = 0.5\, t_1$. That is,

$$0.5\, t_1 = \sqrt{\frac{x^2}{v^2} + t_1^2} - t_1, \tag{42}$$

so

$$1.5^2\, t_1^2 = \frac{x^2}{v^2} + t_1^2, \tag{43}$$

or

$$x^2 = 1.25\, t_1^2\, v^2. \tag{44}$$

Thus,

$$x_{max} = \sqrt{1.25}\, t_1\, v, \text{ where } x_{max} = \text{ the maximum far offset.} \tag{45}$$

The near-offset distance should be short enough to record reflections rather than refractions from shallow events (Figure 158). Generally,

$$x'_{max} = \frac{Z_n V_1}{\sqrt{(V_2^2 - V_1^2)}}, \tag{46}$$

where x'_{max} = maximum near-offset distance, V_1 = interval velocity above a shallow target, V_2 = interval velocity below a *shallow* target, and Z_n = the depth of a shallow target.

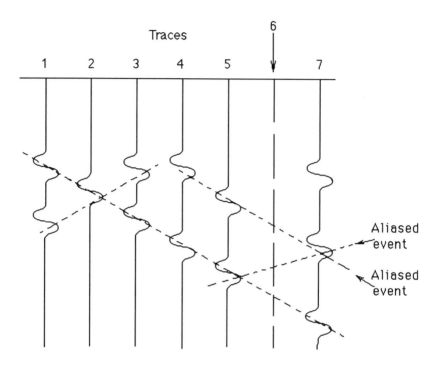

Fig. 159. Stacked section trace aliasing. The addition of a trace at station 6 would define the dip direction.

The minimum near-offset distance should be long enough to ensure that the shot-generated noise level is acceptable. During marine surveys, cable jerk, air-gun bubbles, water turbulence, and ship-propeller noise can cause excessive near-trace noise. With land work, the shortest offset tends to be one station length (about 25 m). In marine operations, it tends to be the distance to the farthest gun from the towing vessel (60–120 m); otherwise, the near receiver would be saturated by gun tow and/or bubble noise.

6.5.2.3   Station Spacing

Receiver stations should be close enough together to avoid the possibility of spatial aliasing. If spatial aliasing occurs on shot records, some transforms (such as f-k) repeat the aliasing in f-k space, so they are no help in reducing coherent noise levels. Spatial aliasing occurs when sampling is inadequate for the frequencies and apparent dips present in the data. For example, spatial aliasing can cause misinterpretation of dipping events (Figure 159). Picking the correct dipping event is just guesswork because the data are aliased.

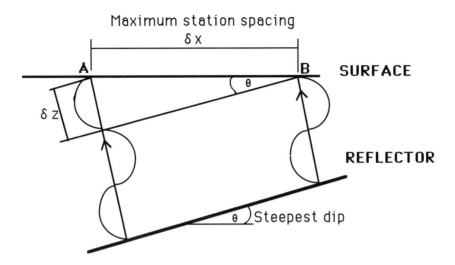

Fig. 160. Dip recording.

In Figure 159, a series of seven CMP traces are shown and trace 6 is missing (indicated with an arrow). Where beds dip and the receiver station distance is too large, spatial aliasing will occur. The missing trace is needed to ensure the correct interpretation because the wavelets on traces 5 and 7 are greater than a half wavelength apart (the interpretation could be that the bed is faulted rather than being continuous). This problem also can occur on shot records. To avoid it, the station interval is set with regard to the desired frequency bandwidth and the steepest apparent dip to be expected. For example, frequencies on the order of 70 Hz require shorter receiver spacing than that needed for 30 Hz data.

The most important parameter is the apparent dip at the receivers. While this depends somewhat on geologic dip, it is more a function of offset and raypaths (for example, at far offsets, a water-bottom reflection can be aliased even if the geological dip is zero). Spatial aliasing also depends on frequency bandwidth of the data. Consider what happens on a shot record when a wavefront from a single dipping bed arrives at two receivers A and B (Figure 160).

Receiver stations A and B are δx apart and the waves arriving at A and B are a half wavelength apart. Aliasing begins when the upgoing received waves are received at A, a half wavelength later than at B, and continues if the waves are farther apart. The additional half wavelength that the reflection travels to A is called δz, where $\delta z = \lambda/2$ (a half wavelength).

Now,

$$\sin\theta = \frac{\delta z}{\delta x} \tag{47}$$

hence

$$\delta x = \frac{\lambda}{2 \sin\theta} = \frac{v}{2f \sin\theta} \tag{48}$$

where v is the medium velocity. If stations were separated by this value of δx or more, then events of velocity v, dip angle θ, and frequency f are spatially aliased. On a shot record in which dip is present and a reflection's moveout increases with offset, the receiver-station spacing should be based on the steepest wavefront arrival dip observed on all stations. To overcome the spatial aliasing problem, we often put receiver stations one-half of the δx distance apart.

That is,

$$\delta x = \frac{v}{4f \sin\theta} \tag{49}$$

where v = average velocity of the shallowest target horizon, f = highest frequency expected, and θ = maximum dip angle.

If there are large elevation changes, high frequencies may be attenuated because of differences in arrival times at each geophone. The δx should then be short enough to allow as small a difference as possible in arrival times from one station to the next.

6.5.2.4   Recording Limitations

The amount of fold is a function of the number of recorded channels and the station-interval/shot-interval ratio (see Chapter 1). Since the station interval is established by the degree of dip and the velocity expected in the area (see above), and the number of channels is limited by the recording instruments, then the fold is determined by the SP interval (Chapter 1). The maximum fold occurs when the shotpoint interval equals half the station interval.

In marine recording, the fold can be limited by the vessel speed, the record length, and how frequently the source can be fired (the source *cycle time*). For example, in subsalt recording where record lengths are longer than normal (around 8–10 s), the cycle time may be as long as 8 s, and if the shotpoint interval is 25 m, then the maximum vessel speed will be 25 m/8 s = 3.1 m/s =

6 knots. In shorter record length operations (such as 6 s), the maximum vessel speed is higher. However, because streamer noise increases as vessel speed increases, the normal operational speed is 4.5 to 5 knots.

The vessel speed often has a bearing on contractors' cost per kilometer because the faster a survey can be shot, the greater the profit that can be made. There is also a minimum vessel speed. A speed of 4 knots or less may cause the vessel (and hence cable) to move erratically (especially under windy conditions), and this may cause source and receiver trailing equipment problems.

Exercise 6.1

Plan a seismic field program to evaluate a permit block of 70 km x 70 km. Seismic data have not been previously recorded on the block, but area geology indicates a favorable prospect of low-relief anticlinal folds with minor faulting. The long fold axes are north-south. The minimum structure considered economically viable to develop has a radius of 2 km; maximum dip is 15°. Table 6.1 shows a sand/shale sequence in an adjacent permit.

Table 6.1. A sand/shale sequence.

AGE	Depth (m)	V_{av} (m/s)	TWT (s)	V_i (m/s)
Recent	300	2000	0.333	2000
Eocene	3000	3000	2.222	3140
Cretaceous	5000	3300	3.000	4137

The shallow horizon will be useful in static corrections. A noise spread gave a prestack signal-to-noise ratio of 0.5:1 at the target depth, but 3:1 is needed for processing. The target frequency is in the range of 15–40 Hz. Hole blowout was avoided if the near-trace offset was no less than 75 m. Table 6.2 shows the crews available.

Table 6.2. Crews available.

Source	Channels	24-fold	km/month	48-fold	km/month
A) Dynamite	96	$300 /km	270	$330 /km	210
B) Vibrator	96	$270 /km	300	$280 /km	230
C) Dynamite	192	$325 /km	180	$355 /km	160

1) Determine the spacing and orientation of a 2-D line grid to cover the area.
2) What CDP fold is required to give a signal-to-noise ratio of 3:1 at the target depth after CDP stack?
3) Determine the best cable geometry: (a) far-channel offset distance to obtain NMO calculations at the Cretaceous and full fold at the Eocene; (b) near-channel offset distance to avoid shot noise but record the Recent-age reflection; (c) maximum channel spacing that will avoid spatial aliasing; (d) minimum number of channels.
4) Compute complete survey length in kilometers.
5) Select a crew from A, B, and C. (a) How long will each take to complete the survey? (b) Determine survey cost for each crew. (c) Determine the shotpoint interval.
6) Briefly note other factors that may influence your choice of the most economical crew. Discuss in general terms crew availability, timing of the survey, budget, government, and the potential to cost-share.

Chapter 7

Three-Dimensional Surveying

7.1 Three-Dimensional Procedures

The goal of seismic data acquisition and processing is to provide a realistic image of the subsurface structure of the Earth. Since that structure is three-dimensional, one might suspect that 2-D data acquisition and processing can be inadequate. That is indeed the case: 2-D seismic data can provide a misleading image of the earth's subsurface, even for the simplest 3-D structures. Figure 161, for example, shows a reflector that has a dip direction perpendicular to the shooting direction of a 2-D seismic line. When the data are processed, the cross dip causes two problems in the image. First, the part of the reflector being imaged is not the part that actually lies vertically under

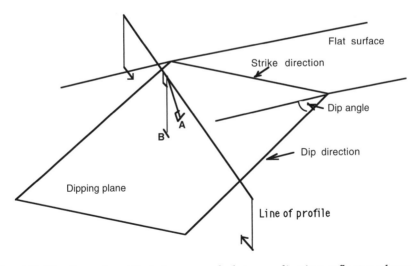

Fig. 161. The line of profile being recorded over a dipping reflector plane. The reflection from point A appears at point B beneath the 2-D line.

the 2-D line, as an interpreter unaware of the cross dip would assume. Second, the reflector's depth is incorrect. Both of these problems occur because the actual reflection points do not lie in the vertical plane below the line.

The solution to these problems is to record 3-D seismic data and to process them using 3-D rather than 2-D imaging algorithms. In 3-D acquisition, the data are collected over a 2-D surface area instead of along 1-D lines. After 3-D processing, reflection events appear as 2-D surfaces rather than as 1-D events. A dipping, reflecting plane such as that in Figure 161 is then correctly imaged.

French (1974) illustrated the benefits of 3-D versus 2-D seismic methods in a classical modeling study performed at Gulf Research Laboratories. One of French's models had a normal fault at the base of the first layer plus a channel and a ridge between the middle and bottom layers (Figure 162). A synthetic seismic profile running at a 45° diagonal across the model was studied.

The synthetic 2-D time section of the diagonal profile is shown in Figure 163a. The reflection from the interface between the upper and middle layers is apparent from 0.5 to 0.8 s. The fault is undefined, but a strong diffraction from the top of the fault is apparent, curving downward toward and overlapping with the upper layer on the left. Because the 2-D line was

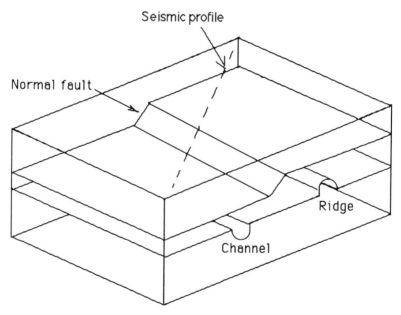

Fig. 162. A model French used to illustrate the effect of 3-D features on 2-D data.

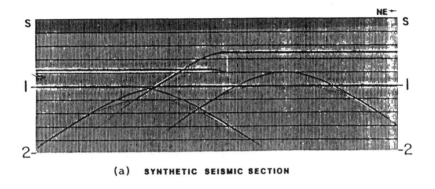

(a)   **SYNTHETIC SEISMIC SECTION**

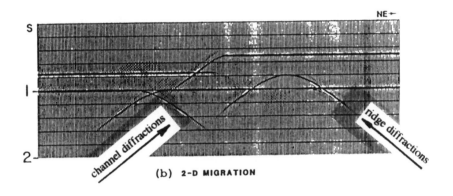

(b)   **2-D MIGRATION**

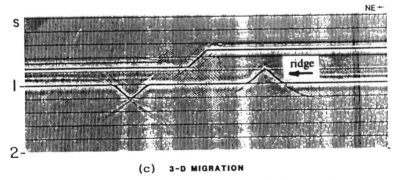

(c)   **3-D MIGRATION**

Fig. 163. (a) Time section from a 2-D line run at 45° across the model of Figure 162. (b) A 2-D migrated line. (c) A 3-D migrated line (after French, 1974).

"recorded" diagonally across the fault, the diffraction pattern is partially out of the plane of the profile. The channel to the left in the model appears as two curved diffractions. These are caused by reflections from both sides of the channel. The ridge is marginally better defined than the channel, though its representation is much broader than the true ridge width. The diffractions from the flanks of the ridge are as strong as those from the channel, curving downward toward the bottom of the section.

Figure 163b shows the section after 2-D migration processing (see Section 7.5 for a discussion of diffractions and migration). The fault is of shorter width, and the channel is becoming imaged in a V shape. The ridge is also narrower, and for all three shapes, the diffractions are shorter in length. This is because the migration process shortens the diffractions by collapsing them to their point of origin. The diffraction data that remain on the section are those data that arrived from the off-line direction. These data cannot be migrated by 2-D processing.

Figure 163c shows the section after 3-D migration processing. Data have been acquired in the off-line direction, migrated into the in-line direction, and then migrated along the line of profile. The fault, the channel, and the ridge are now positioned correctly and are imaged better. Diffraction patterns still exist as a result of either the migration processing not being perfect or portions of the diffractions arriving from directions other than that of the migrated off-line direction.

Historically, 2-D surveys were the main method for imaging geology. More expensive 3-D surveys were used only when detailed geologic information was needed during reservoir development. However, the ability to correctly image a 3-D target using 3-D rather than 2-D acquisition and processing techniques, as shown above, is a powerful argument for recording 3-D data from the outset. In well-developed areas such as the United States, this argument has been accepted. Three-dimensional techniques generally are adopted from the outset because the cost of a single 3-D survey that images the geology properly is less than that of a series of 2-D surveys performed over a period of time (Nestvold, 1992).

Three-dimensional seismic data, properly recorded and processed, represents a 3-D image of the subsurface geology, sometimes referred to as a "volume of the Earth." Using such data, an interpreter can (with computer aid) reconstruct 2-D profiles along any direction or azimuth. Figures 164 and 165 show this concept in isometric and plan views. In a 3-D data volume, the traces contributing to the stack at each CMP location or *bin* can have both different offsets and different azimuths (see Section 7.5). Three-dimensional processing steps account for this so that each bin stacks properly. If a desired 2-D profile does not intersect bin centers, then trace interpolation can provide an evenly sampled profile.

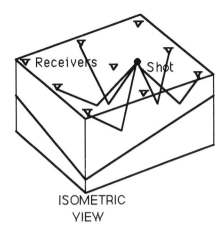

Fig. 164. Three-dimensional recording in which a single shot is fired into many receivers, located at all azimuths. The reflection raypaths occur at all azimuths, dependent on the location of the receivers.

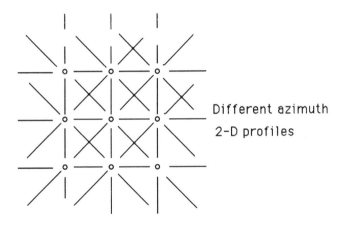

Fig. 165. Plan view of different azimuth 2-D profiles that can be obtained from a 3-D volume.

Two-dimensional profiles selected from a 3-D data volume do not have to be vertically oriented planes but can be oriented in any direction. One popular interpretation tool is the *time slice*, a horizontally oriented 2-D profile. This kind of profile gets its name from the vertical axis of a 3-D volume, which is often time. The name indicates that the profile cuts across the data volume at a constant time.

The interpretational power of time slices is shown in Figure 166, which depicts a *listric fault* (a fault that flattens as it increases in depth) as it "tracks" across four time slices. As time increases, the fault boundary moves northward as indicated by the arrows. At 2 s, most of the fault boundary is no longer within the 3-D data volume. Time slice interpretation often is done on computer workstations where the ability to rapidly scan through a sequence of slices helps the interpreter visualize the 3-D character of a reservoir.

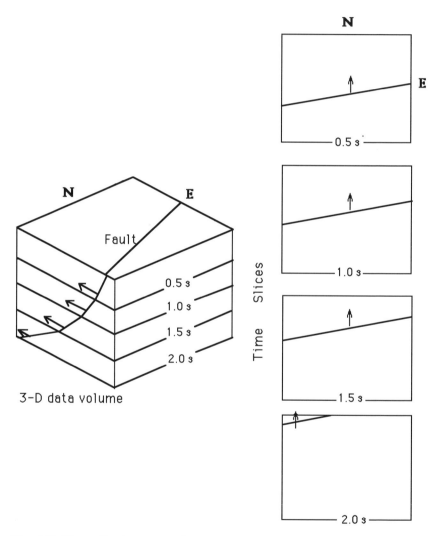

Fig. 166. Time slice interpretation.

As was discussed in Chapter 1, the cost of acquiring 2-D seismic data per kilometer offshore is less than the cost of acquiring that same data on land. This statement equally holds true for 3-D surveying. However, the techniques of acquiring 3-D data differ for land and marine surveying. A ship tows one or more seismic cables behind it, and this configuration cannot be changed easily, apart from the potential to use two vessels at the same time. The land configuration, by contrast, can be much more versatile. Since reflected seismic waves arrive at the surface from all azimuths, it is beneficial to record seismic data at all azimuths. The land seismic technique can accomplish this, but the standard marine technique cannot. The two approaches will now be discussed, beginning with marine 3-D, the simpler of the two.

7.2   Three-Dimensional Marine Surveying Method

In marine 3-D seismic surveys, the geometrical constraints imposed by using a towed streamer allow only a few types of survey designs to be practical. Most 3-D marine surveys recorded with towed streamers are designed as a sequence of closely spaced (say 50–150 m) parallel 2-D lines. (One alternative, circle shooting, is described in Section 7.4. Marine 3-D recording using bottom cables also is described in that section.) The parallel 2-D lines provide areal CMP coverage that, after processing, gives a 3-D image of the subsurface. Except for the tight line spacing, this type of marine 3-D acquisition is similar to 2-D acquisition. The sequence of 2-D lines can be shot in several ways; Figure 167 shows one such pattern, in which the shooting direction reverses every line. Other *sail lines* patterns are possible. Generally, a marine crew selects a pattern that minimizes the time spent moving from the end of one line to the start of the next.

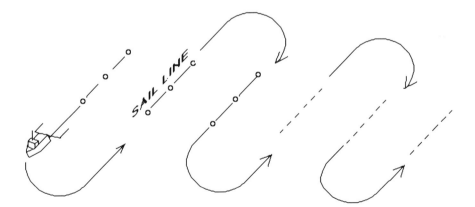

Fig. 167. Marine 3-D recording technique.

During the early days of recording marine 3-D surveys, data were recorded using a single vessel, a single streamer, and several air-gun strings acting as a single energy source. This meant that each traverse of the survey area by the sail line produced one line of subsurface coverage. A typical early (1970s) survey had parallel lines about 10 km long, spaced some 50 m apart. If the seismic vessel towed the streamer at 5 knots, then each line would take just over one hour to shoot. Because the vessel turning time between lines was also about an hour, on such surveys the vessel was productive for only half the time. Consequently, contractor service companies preferred to bid for seismic surveys on a time rate or daily rate, rather than on a kilometer ("turn-key") basis. Many early surveys were recorded and processed by the same contractor because a convenient "package" cost for acquisition plus processing could reduce the overall cost to the client exploration company.

Because the cost of 3-D marine acquisition was so high, during the 1980s new ideas were considered to increase the speed of data acquisition, thereby lowering costs. One idea was to record data using two well-coordinated ships sailing side-by-side, each towing a streamer and an air-gun array. The sources were fired in an alternating sequence, while data were recorded by both streamers for every shot. In this fashion, three seismic lines were collected for the price of two. That is, each ship recorded a standard line plus a line covering CMPs halfway between the two vessels. This acquisition configuration also allowed subsurface coverage to be obtained under obstructions such as producing platforms (see Section 7.4).

Economics is the driving force behind the technological advances in 3-D marine acquisition. The company with crews that can collect the most quality data at the lowest cost will get the most business. If a ship tows two cables rather than one, its production rate almost doubles, with a much lower percentage increase in costs. Consequently, during the late 1980s, contractors started to tow a number of streamers and sources from a single vessel to increase productivity. With two sources in the water, it was possible to fire them separately and record data separately on the two streamers. The ship power to tow two such streamers would render the conventional seismic vessel (which was often little more than a modified rig supply tender) as inadequately powered. Furthermore, towing two streamers (known as *dual-streamer* operations) and air-gun arrays required wider back-deck space and greater air compressor power.

The result was the commissioning of so-called "super ships" by contractors such as Western Geophysical and Geco-Prakla. An example of a ship towing three streamers and two gun arrays is shown in Figure 168. If gun array 1 fires first, then the vessel would record data from CMP line 1 at streamer 1, CMP line 2 at streamer 2, and CMP line 3 at streamer 3. When gun array 2 fires, data of CMP line 2 are recorded at streamer 1, CMP line 3 at streamer 2,

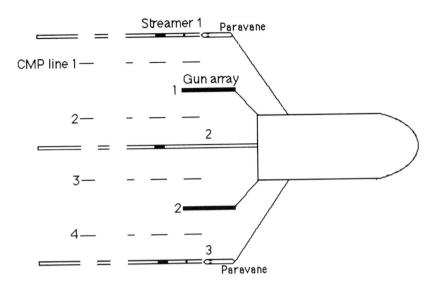

Fig. 168. A multisource/streamer configuration.

and CMP line 4 at streamer 3. By firing the two-gun arrays alternatively every 25 m (i.e., individually every 50 m), four CMP lines may be recorded by the two-gun, three-streamer configuration shown here. Alternatively, by changing the cross-line offsets of the sources or streamers, it is possible to record six CMP lines at once.

To position streamers and sources outboard of the vessel and separate from each other, two approaches are used. One approach is to tow the trailing equipment from booms swung out from the vessel's sides. The other approach is to tow equipment from paravanes (or "baravanes"—a Halliburton trade name). A paravane is a steel chamber, shaped like a cigar; that houses remotely controlled fins (controlled from the vessel) and other mechanisms to allow it to move sideways and dive to a required depth, if necessary. Paravanes also may house transponders for navigational positioning.

With such super ships, more data could be recorded in one day than had been dreamed of only 15 years previously. For example, with a single streamer and a single source, a vessel shooting a continuous line 24 hours per day could not record more than 200 km. Allowing for mechanical and other technical failures, the record for such 2-D work—some 180 km in 24 hours—was held by a vessel recording in the Arctic Ocean in 1987. A multistreamers configuration using four streamers changed the book of records to 1200 km in a 24-hour period in 1992. This record lasted only 18 months, after which new records were set by super ships that could tow as many as six to eight cables.

A recently launched ship has been built to tow as many as 12 cables. The economic incentives to increase productivity probably never will disappear. Consequently, further technological advances that lower the cost per unit of 3-D coverage are likely.

Whether a seismic ship is towing a single streamer and source or many streamers and sources, the positions of the towed systems are affected by winds and currents. Figure 169 shows a phenomenon called *streamer feathering*, which occurs when there is a current having a component in the cross-line direction. Feathering introduces a cross-line component to CMP positions. During data processing the location of each trace's CMP must be known so it can be assigned to the correct stack bin. Because of feathering, the actual subsurface coverage obtained by one traverse of a survey area is seldom the same as the planned coverage. Thus, accurate source and receiver positioning data must be recorded and processed during data acquisition to ensure that the actual subsurface coverage meets the survey coverage specifications.

When only a single streamer and source were towed, the positioning equipment and processing systems were quite modest. Typically, a streamer would contain four to 10 compasses whose data would be integrated to reconstruct the streamer shape. The tail-end position of the streamer was

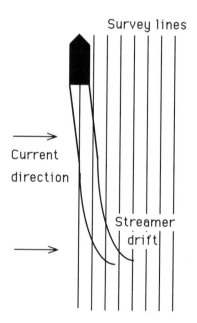

Fig. 169. Streamer drift can cause midpoints to be located off-line.

monitored by the ship's radar. The front section of the streamer and the source were located using acoustic triangulation measurements. Some crews used *tow sensors* to measure the angle at which the streamer left the ship. All of these data were processed in real time to provide a continuous monitoring of subsurface coverage.

With the advent of ships towing several streamers and sources, the positioning systems became more elaborate. Figure 170 shows an example. Typically, the near-offset receiver and source positions are determined by a system of transponder pingers and receivers. Each such pair provides an acoustic range measurement of the distance separating the pair. Many such measurements can be combined to determine accurate positions, just like in the range-range ship-navigation systems described in Chapter 5. Acoustic systems are often also deployed at the tail end of the towed streamers and sometimes at a middle offset. GPS receivers and laser range finders may be positioned on streamer tail buoys and other buoys to provide additional positional data. All of the data together make up a so-called *positional network*. The network data are inverted in real time by powerful workstation-class computers to provide accurate positions for all of the sources, receivers, and midpoints. A CMP coverage map is maintained by the computer so that any coverage shortcomings can be seen and subsequently fixed by shooting *in-fill lines*. Although required positional accuracy is dependent on CMP bin size, current industry practice is to aim always for average positional errors of 5 m or less.

In some areas, such as the North Sea, changing and unpredictable winds and currents cause the initial CMP coverage to have many holes. Sometimes as much as 30% of data acquisition time is spent shooting in-fill lines to correct coverage deficiencies. Survey budgets should allow for such contingencies in areas where they are likely to occur.

7.3 Three-Dimensional Land Surveying Method

In 3-D land recording, there are a number of source/receiver configurations that may be used. Ideally, we wish to produce a gather of data containing all azimuths when feasible (because if the raypath azimuths are from all directions, then the data are truly three-dimensional). To do this properly, the source/receiver lines may be positioned at right angles to each other, as shown in Figure 171. This configuration is commonly known as the *crossed-array approach*, in which the source is fired along the source line toward the receiver line as a broadside shot, eventually crossing the receiver line in split-spread manner, then continues firing as it moves away from the receiver spread. The shot records commence with the reflected waves arriving broadside, becoming progressively hyperbolic until in the split-spread configuration, when they appear like normal split-spread shot records before becoming

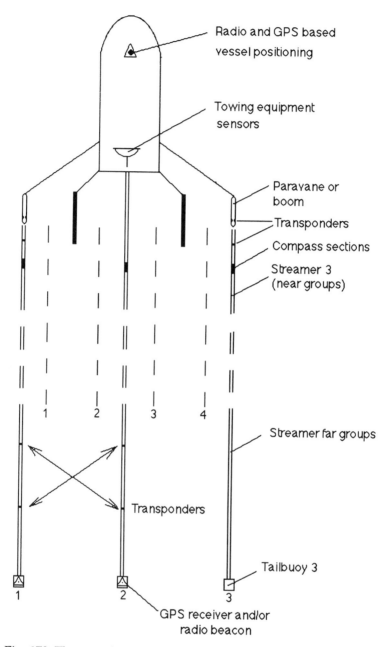

Fig. 170. The use of multiple streamers and sources requires cross-communication transponders.

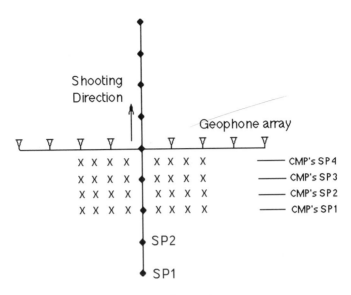

Fig. 171. Crossed-array configuration.

more broadside as the source moves away from the receiver line again. The CMP lines are shown for each shot in Figure 171, where a full crossed-array recording spread produces a set of single-fold traces with a total areal size one fourth that of the crossed-array area.

The benefit of using the crossed-array approach is that all azimuths are recorded, which is the ideal situation for 3-D acquisition. However, such benefits are often outweighed by the realities of ground roll and geologic dip. Ground roll is prevalent in land seismic recording and masks useful data as discussed in Chapter 2. It can be removed by the use of source and receiver arrays (Chapter 2), by the "stack-array" technique, or by the use of *f-k* processing techniques (Chapter 6). All of these methods of ground-roll reduction rely on the fact that the ground roll has a relatively constant apparent wavelength and/or velocity of propagation.

When the crossed-array source/receiver configuration is used, the ground roll not only changes apparent wavelength but also velocity as the source changes azimuth with respect to the receiver line. Rapid changes in apparent wavelength or velocity often render transforms such as *f-k* unusable. The use of source/receiver arrays also fails unless the arrays are three-dimensional and spread out on the ground in star shapes or other such patterns (Sheriff, 1991a). In practice, such arrays are not used because the short station distances make the arrays cumbersome (and therefore time consuming and expensive) to handle. It is preferred to leave array forming or other ground-roll cancellation transforms to the processing center. Consequently, with the crossed-array technique for recording 3-D data, there is no conventional method for removing ground roll in the field or in the processing center.

However, the crossed-array approach does have a further benefit. Since source lines run orthogonally across the center of the receiver lines, it can ease access problems in areas of country where there are rectangular cross-tracks, such as farms with ploughed paddocks. Low-fold data can be recorded under such conditions when access problems exist. This approach to recording low-fold 3-D data has been termed the *lofold3d* technique by Allied Geophysical Laboratories (University of Houston), which pioneered the approach in the mid-1980s. This was prior to the publication of the results using the technique by Allen and Howell (1990), who termed the technique "poor man's 3-D."

To build up fold in the lofold3d technique, data are recorded over one area, then the receiver line is moved along half a receiver line and the next source line is shot across the center of the next receiver line, as shown in Figure 172. As each receiver line is crossed, two-fold data are built up by the overlapping bins. After completing the desired strip of two-fold data (with single fold at the ends), the receiver line may move up a half source-line length and the operation repeated. This produces three-fold coverage, considered as low by any standard. However, provided the receiver stations are closely spaced along what tracks are available, different sized CMP bins may

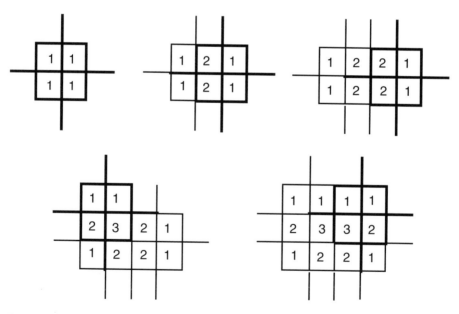

Fig. 172. Building fold with the crossed array.

now be mixed so that if four bins close together (each of four-fold data) are mixed, they will produce a 16-fold bin. This would be typical of the level of fold in a land 3-D survey, the fold never attaining the high level of fold prevalent in marine surveying.

Another source/receiver line configuration is *swath recording*, which uses parallel receiver lines and allows the source to move up and down the line close to the receivers as shown in Figure 173. CMPs are again located halfway between each source shotpoint and each receiver station.

In Figure 173, 12 receiver stations (open circles) are positioned along each parallel receiver line so that if a 120-channel recording system were used, there would be 10 receiver lines. Seven receiver lines are shown. If they were all active, they all would receive a number of different azimuth data. However, in this case, since the parallel receiver lines are closely spaced compared with their length (making the cable layout more like a strip than a square), the majority of offsets are biased toward a minimum (in-line) azimuth with most of the data being recorded in-line rather than broadside, as it is with the crossed-array approach. Therefore, with this approach, noise-canceling arrays operate better, and even 2-D DMO processing has been used successfully as opposed to the more expensive 3-D DMO processing (Lansley, 1992).

Swath recording is the most popular form of recording 3-D data, but it requires virtually unlimited source and receiver access to the area to be sur-

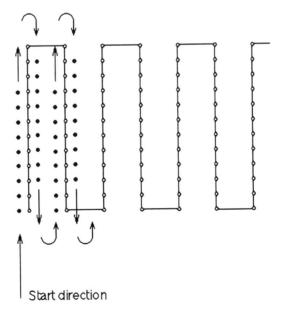

Fig. 173. Swath recording.

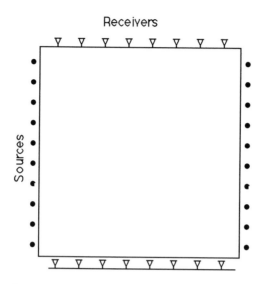

Fig. 174. Framework 3-D recording.

veyed. There must be easy access for the energy source along each side of each receiver line. Vibrators are the preferred energy source for this form of surveying.

In farming areas, it is difficult for seismic survey crews to gain entry to work (either because the farmer demands a cash payment for entry or because there are cultivated farmlands within the region to be surveyed). Many farming areas are located on one-mile rectangular grids, with public access no closer than along the outskirts of the rectangular grid. The only way seismic data can be recorded over such areas is by undershooting them. This is *framework* (or "seisloop") recording and is performed by placing the geophone stations along two of the rectangle's parallel sides and using the other two sides as source lines, as shown in Figure 174. After recording the data with this configuration, the source/receiver perimeter lines may be swapped. If the recording system has a high recording-channel capacity, it may be possible to completely encompass the area to be surveyed with receiver stations and run the source around the perimeter. Either way, the fold of coverage is not as high as that using swath or crossed-array techniques.

7.4 Other Marine Survey Methods

When performing transition-zone surveys where the water is extremely shallow, hydrophones can be attached to buoys and placed on the seabed. Alternatively, ocean-bottom cables that are not buoyant may be laid on the seabed. Both of these methods allow a source vessel to run along the receiver lines (as in the land-swath recording technique shown in Figure 173), or across the receiver lines (as in the crossed-array land recording technique shown in Figure 171). In the early years of recording marine data, modified land cables were deployed on the seabed, the action often being referred to as back-down operations because it required the recording vessel to take the cable to a location, drop the cable to the seabed, and then back down the vessel when retrieving it. The shot could be fired either by the recording vessel or a separate shooting vessel.

7.4.1 Circular Shooting

The point was raised earlier that line-change time reduces the efficiency of marine 3-D operations. French, while employed at Grant Tensor, proposed that instead of shooting along straight lines to collect the required CMP coverage, data could be recorded in a circular manner (which was thereafter referred to as *circular shooting*), thereby reducing the line-change time to zero. Figure 175 is an example of how an area may be recorded using circle shooting. Note that the bins are filled with different azimuth data (which some consider better for imaging 3-D geology than the conventional linear approach).

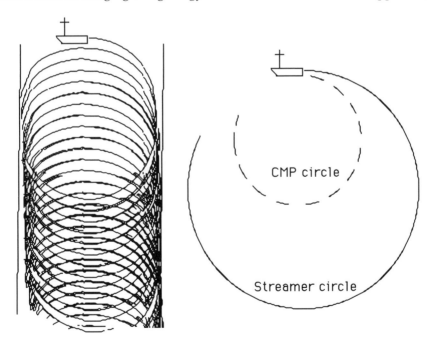

Fig. 175. One possible circular shooting sail line.

However, a simple sketch of the approach shows that a much greater fold occurs at the outer edges of the volume than at the inner points. Since the extra fold does not generally provide any significant information, the time spent recording them is an inefficient period, analogous to line-change time. Alternatively, the ship can sail a spiral pattern, thereby avoiding the recording of redundant traces. The problem this causes is that a greater noise is evident on front-end groups than is normal. Onboard navigation systems must be closely monitored, not just to ensure correctly filled bins but also to ensure that the streamer(s) do not come in close proximity to the shooting vessel. (Such a dangerous situation can happen rapidly if the current direction changes dramatically.) The postprocessing of these navigational data is more complex than that for the normal linear configuration of recording 3-D, and the exploration company must weigh the advantage of a reduction in sail-time costs compared with an increase in navigation processing costs.

One area where the circular seismic method has been acclaimed is shooting around a salt dome. With such structures, recording a seismic line over the top of the dome can be a waste of time since the important oil and gas traps lay on the flanks of the dome. Consequently, circular seismic surveying has been successful in recording data around a dome in a circular manner. The circular sail-line grid may be followed by a few sail lines crossing them to tie in the data from the circular lines. This is to ensure that the navigational and seismic data have been recorded correctly and tie well. Any mis-ties (such as events that do not arrive at the same two-way time on the seismic section) alert the processor to either seismic or navigational problems. The data are binned in the conventional 3-D manner, and azimuths are much more variable than those for a linear survey. The survey, consisting of circular and straight 2-D lines, is shown schematically in Figure 176.

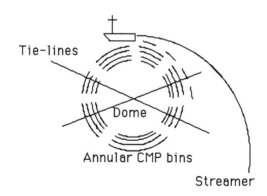

Fig. 176. CMP seismic lines over and around a domed structure.

The series of circular lines complete an annular grid of lines around dome structures, and the line separation is dependent upon the geology to be imaged, as is normal practice. All of the conventional 3-D navigation technology is required here, so the survey cost is similar to that of a normal 3-D survey, with the exception that there are very few line tails.

7.4.2 Two-Vessel Operations

Many marine 3-D surveys are shot in mature producing areas where permanent structures such as production platforms make it impossible to get uniform CMP coverage with a standard survey. Figure 177 shows one solution to this problem: two-vessel acquisition. Seismic vessel A may be only a source boat, whereas vessel B tows only a streamer. As they sail on opposite sides of the obstruction, CMP coverage is obtained under the obstruction. Alternatively (as discussed in Section 7.2), both ships can have sources and streamers so that three lines of coverage are obtained at once. In either configuration, the source firing times and recording start times must be synchronized between the two vessels. This is accomplished by a continuous radio link between the ships.

Before recording such data, it must be ensured that feathering will not cause the streamer to stray near platform chains or buoy chains (which encircle the platform). Therefore, a current meter is often positioned near the platform or rig to provide information on the local current direction. (Although the seismic crew monitors current by checking streamer position, many pro-

PLAN VIEW

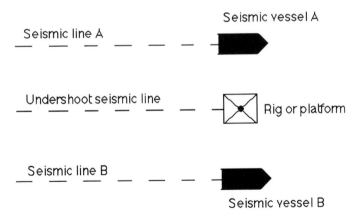

Fig. 177. Two vessels undershooting a platform.

duction companies still prefer to monitor the current status directly using a current meter, since this also can provide other useful local information.) A current meter is basically a paddle wheel that turns to the direction of maximum current (like a weather vane) and transmits both current intensity and its predominant direction, via radio, to the seismic vessel. In the event that currents pose a problem, either the ships move off to work elsewhere or they configure their towing direction to ensure the cable feather is away from the chains.

7.4.3 Reconnaissance Surveying (3-D or 2.5-D Surveys)

Reconnaissance 3-D surveys are not full comprehensive 3-D surveys and are, therefore, often sarcastically referred to as "2.5-D surveys." Such surveys lower the cost of full 3-D data acquisition and processing by recording every second or third line of a conventional 3-D survey. Lines between the recorded data are interpolated in the processing center. While such surveying is acceptable in areas of simple geology, it will often miss fine featured geology such as faulting. In zones of complex faulted geology, a reconnaissance survey is unacceptable and leads only to erroneous interpretation. Consequently, such "recon 3-D" surveys are generally frowned upon by the industry and often are viewed as penny-pinching because the overall cost of field development is much larger than the cost difference between a true 3-D survey and a recon 3-D.

7.5   Three-Dimensional Survey Design

Because of their greater complexity and cost, 3-D surveys are more difficult to design than 2-D surveys. The primary goal of 3-D survey design is to ensure that field acquisition parameters and processing parameters produce an interpretable image of the exploration target. An important secondary goal, however, is to minimize the acquisition and processing costs of the survey. Unfortunately, these two goals are usually at odds with one another. To satisfy both of them, a survey designer must consider not only the purely geophysical issues but also issues such as crew logistics, survey-area accessibility, weather, personnel costs, time sharing with other crews, and so on. All of these complications make an exhaustive treatise on 3-D survey design beyond the scope of this book. Instead, this section concentrates on only the geophysical parameters that affect the processed image. A more in-depth study of 3-D survey design is given by Stone (1994).

Design of a 3-D survey begins by specifying the depth and areal extent of the target image and the largest dip among events that will contribute to the final image. The maximum dip is an important geophysical parameter for two reasons. First, it governs the spatial sampling of a survey. That is, sources, receivers, and CMPs must be spaced so that spatial aliasing is not a problem in the steep-dip events. Second, steeply dipping events from within the target area can produce their reflected energy at the Earth's surface in regions beyond the target edges. If such energy is not recorded, then migration processing cannot refocus that energy back into the target image. Thus, the amount of dip in an area ultimately determines the physical size of a survey needed to image the area.

7.5.1 Sampling

An adequate spatial sampling is necessary for the dips and frequencies expected; otherwise, spatial aliasing will occur. Temporal aliasing also will occur if inadequate temporal sampling is performed. Depending on the geology to be imaged three-dimensionally, a typical 3-D survey has CMP lines anywhere from 25 m apart in marine recording to 100 m apart in land recording. Spatial sampling is now a 2-D rather than a 1-D issue.

As explained in Chapter 6, in practice, spatial aliasing is avoided by making the receiver separation Δx, either $V/3f \sin \theta$ or $V/4f \sin \theta$. For example, using $\Delta x = V/4f \sin \theta$, and if $V = 3000$ m/s, $f = 20$ Hz, $\sin \theta = 0.5$, then Δx should be less than 75 m. This applies for in-line receivers as well as cross-line receivers, and so determines the 3-D line-separation distance. Thus, we ideally prefer ΔL less than $V/4f \sin \theta$, where ΔL is the line spacing. In computing Δx, the survey designer should use for f the highest nonspatially aliased frequency desired. For a review of temporal sampling, see Chapters 2 and 4. If a target horizon has $V = 3000$ m/s, $\theta = 10°$ maximum, and a frequency of 60 Hz, then the line spacing,

$$\Delta L = \frac{3000}{4 \times 60 \sin 10°} = 72 \text{ m.} \tag{50}$$

If $\theta = 15°$, then $\Delta L = 50$ m; or if $\theta = 5°$, then $\Delta L = 143$ m. A line spacing of 50 m is considered fairly detailed for a marine 3-D survey, while a line spacing of 143 m would be similar to that adopted by marine reconnaissance 3-D surveying. By contrast, the larger line separation would be typical of land 3-D. The reason for this difference is that a land survey costs more than a marine survey, and so line spacing, in practice, can become more a question of economics than obtaining the best-sampled 3-D data.

7.5.2 CMP Binning

A *CMP bin* is the name for an area that will provide a collection of seismic traces to be stacked in 3-D data processing. Bin sizes generally are determined by the signal-to-noise ratio and the CMP spacing. Since the CMP (and line separation) spacing is decided by a knowledge of the dip, frequency, and velocity values, then the reflection signal-to-noise ratio and resolution (both horizontal and vertical) become the most important factors in determining bin size.

The signal-to-noise ratio in marine surveying is generally of a relatively high value because data are collected with a high fold of coverage, similar to that of 2-D surveys. However, in land 3-D recording, establishing the minimum signal-to-noise ratio and resolution becomes a problem because, often, land recording does not have high coverage and dense spatial sampling. Thus, ways of enhancing the signal-to-noise ratio after the data acquisition are necessary.

A common practice is a processing step that takes the desired traces from adjacent bins by changing the bin size. This method is called "dynamic" or "elastic" binning, as opposed to "static" binning (where bin sizes have been set by the CMP and line spacing). For example, if the signal-to-noise ratio is not as good as is expected in the strike direction, it is feasible to take pairs of bins in the strike direction and stack the traces as elongated bins (one bin width in the dip direction, two bin widths in the strike direction). While this may improve the signal-to-noise ratio and reflector continuity in the strike direction, the final stacked section may deteriorate since this process takes data from two adjacent CMP lines. Dip moveout processing goes some way toward resolving this issue. Nevertheless, such elastic binning is treated with caution and is commonly performed to overcome minor coverage imperfections, where spatial resolution and reflector frequency content allow such binning.

Each bin ideally should have a minimum fold amount, a range of offsets, and a range of azimuths. While the minimum fold amount is not a problem for marine surveying, it can become a problem with land 3-D where recording over or through surface obstacles or rough terrain can be difficult. Having a full range of offsets can become a major problem for both land and marine 3-D. In land 3-D, recording a complete range of offsets may be difficult because of the same problems as those limiting the fold amount. In marine surveying, streamer feathering may cause erratic offset binning as the longer-offset trace data stray into adjacent bins.

A full range of azimuths is the most difficult to bin successfully in marine 3-D because the recording configuration in marine seismic is mainly off-end. This is not the case in land operations, where a full range of azimuths is possible, as the source can move around the receiver cable to all desirable azimuths. Figure 178 is a typical azimuth-versus-range quality check diagram that is often constructed before a land seismic survey.

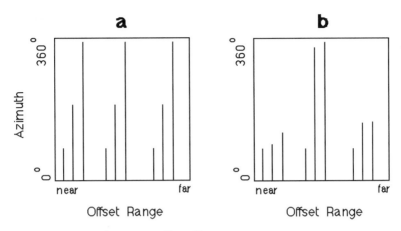

Fig. 178. Azimuth versus offset diagrams.

In Figure 178a, the azimuth range is broad over a group of near-, medium-, and far-offset distances for a single CMP bin. This is considered ideal. However, the situation often exists, as shown in Figure 178b, where the offset groups have erratic azimuths. Such an azimuth/offset range is not desirable in land surveying, and decisions must be made to either improve the azimuth range or find an alternative solution (such as elastic binning).

Generally, such diagrams are not used before recording a marine 3-D survey because of the minimum range of azimuths possible (in marine 3-D, in which the near groups are closest and almost parallel with the gun arrays compared with the almost in-line far groups, the largest azimuth is that from the source to the near offset group). Seabed and transition-zone 3-D surveys, in which the receiver cable is placed on the seabed or shore and the seismic vessel runs around the stationary cable, are the closest that the marine survey ever comes to recording a large range of azimuths.

7.5.3 Migration

A major step in 3-D survey design is deciding the extent of surface coverage needed to obtain a desired subsurface image. When geologic layering is flat, reflection points lie at the CMP point, midway between the seismic source and the receiver. When geologic layering is not flat, reflection points no longer coincide with the CMP. This can be visualized easily using the geometric construction of a reflector's "source image point." To explain this geometrically, we can consider the reflecting plane to be like a mirror so that we can draw lines to an image on the other side of the mirror. The line from a mirror's image to the receiver results in the reflection point being directly between the midpoint between source and receiver. With seismic rays, the straight line to the receiver from the image point shows the travel path of the seismic ray from source to reflection point, to receiver, as shown in Figure 179a.

If geologic dip of an angle "θ" is now introduced, the reflection point moves updip as shown in Figure 179b and is no longer beneath the midpoint between source and receiver. To return these data to the correct midpoint, we perform a processing technique called "migration." Migration is "an inversion operation involving rearrangement of seismic information elements so that reflections and diffractions are plotted at their true locations" (Sheriff, 1991b). Migration takes the data caused by diffractions (see Chapter 1) and returns them to the CMP from which they emanated. This is known as "collapsing the diffractions" and is a major processing step usually applied to the seismic section after stacking. Event movement by migration must be allowed for when establishing the size of a 3-D survey needed to image a particular reflection area.

When planning a survey to generate a seismic volume consisting of CMP traces, two issues are important. First, there must be additional coverage at the edges of the 3-D volume to ensure that the target image has full-fold cov-

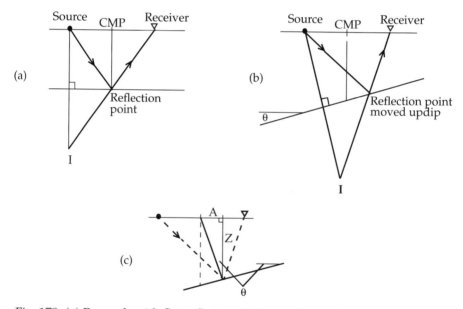

Fig. 179. (a) Raypath with flat reflector. (b) Raypath moving updip. (c) Aperture A, the additional distance to be recorded to provide data for migration processing.

erage across the area intended to be mapped. Second, there must be additional coverage because of the behavior of migration processing. At the end of seismic lines, therefore, a tail end of data (half the streamer length) is recorded to obtain a stacked full-fold coverage at the desired last full-fold CMP. We also must record additional lines for migration processing where necessary.

Because reflection points move when dip is present, we must record a distance A, called the migration aperture, past the end of the line.

In Figure 179c, the reflection point has moved updip as a result of dip angle θ. In the right-angle triangle constructed in Figure 175c with sides A, depth Z, and acute angle θ,

$$\tan\theta \; = \; \frac{A}{Z}. \tag{51}$$

That is, the aperture

$$A \; = \; Z\tan\theta. \tag{52}$$

Consequently, each CMP seismic line must be increased at each end by the aperture length to allow for migration processing along the line. This extra length must be accounted for in the data-acquisition budget. Because there are two apertures (one at each end of a line), the total amount to be added to a line sometimes is referred to as the *aperture* and the half at each end as the *half-aperture*. Whatever terminology is used, the most important point is that an aperture exists at each end of a line, which must be added for migration processing. This is as applicable to 2-D recording as it is to 3-D.

If deep steeply dipping reflections are to be imaged and correctly migrated, the aperture can add a substantial length to each line. Geologic dip is three-dimensional, which requires 3-D migration processing. This means that an aperture is required all around the seismic program area, like a window frame as shown in Figure 180.

The aperture width can be different in different directions. For example, in the dip direction, the aperture will be wider than in the strike direction. In Figure 180, the aperture A in the dip direction will be greater in size than the aperture a in the strike direction because the dip direction has a greater value for θ.

In most seismic surveys, the major lines of interpretational interest are in the dip direction (see Chapter 1), and, therefore, most seismic lines recorded in a parallel direction in a 3-D survey are recorded in the dip direction. Where there are large velocity contrasts across the shallow section of faults, it has become a practice to record 3-D surveys in the strike direction, which avoids

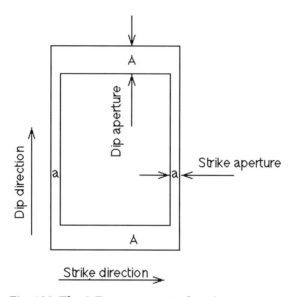

Fig. 180. The 3-D aperture window frame.

the possibility of 3-D data processing stacking-in false back faults (Evans et. al, 1995). Since 3-D surveys require an aperture around the outside of the program area, not only must the dip lines be increased in length to allow for the dip direction aperture but more lines must be added in the strike aperture.

Exercise 7.1

1) For the 28-trace swath example shown in a plan view in Figure 181, the shotpoint locations are marked as "x" and receiver stations by triangles labeled R1, R2, etc. This configuration is possible during normal land 3-D operations or marine bottom-cable 3-D operations. Firing SP1 and SP2 into receiver line 1 generates the single-fold CMP lines as shown schematically on the right (like a stacking diagram on its side), and this continues for eight shots. Indicate where the shots are located on the figure to produce the CMP lines shown, and determine what the maximum in-line fold would be if only source line 1 were fired into receiver line 2.

2) In-line and cross-line fold accumulates when such shots are fired in swath manner, into the full 28 stations. In-line fold increases as shots are recorded along a line as in 1) above, but cross-line fold accumulates as a number of source lines are fired into the same receiver spread. Source line 2 would commence after SP8 when the source

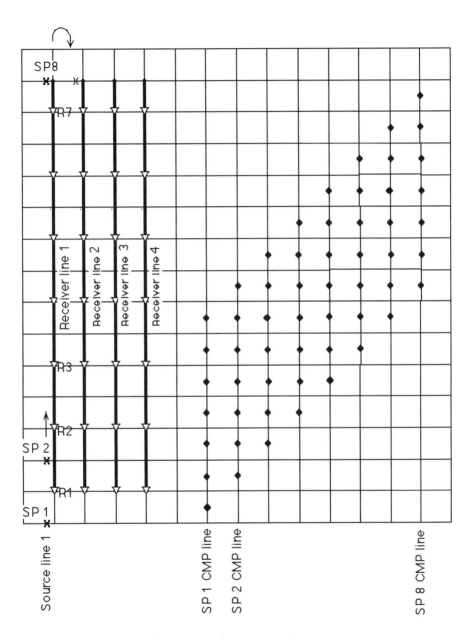

Fig. 181. A 28-trace swath example shown in a plan view.

moves to the next line (as shown by the arrow). Determine for this 28-station receiver spread the maximum cross-line fold, after the source shoots along all four receiver lines.

3) The maximum fold for this recording configuration is the product of the in-line and cross-line fold. Determine the maximum fold.

Exercise 7.2

Plan a 3-D marine seismic survey to cover the prospect area shown on the contour map (Figure 182), using the conventional towed streamer CMP recording technique. The desired frequency spectrum for reflections from the target horizon (top of the contoured structure) is 15 to 40 Hz. Average velocity at the target is 3000 m/s.

1) Determine the maximum dip considering the major features.
2) What should the line orientation be?
3) What digital sample rate should be used to avoid temporal aliasing? What channel spacing should be used to avoid spatial aliasing in the dip direction? What line spacing should be used to avoid spatial aliasing in the strike direction?
4) If a 2400 m streamer cable is used, what is the tail-spread distance required to obtain full-fold CMP over the end of each line marked on the base map?
5) For accurate data processing, calculate the aperture distance A that must be recorded at the end of each line. This will be in addition to the tail spread calculated in 4) above.
6) Calculate the total number of line kilometers required for the survey, taking into consideration the basic line coverage needed to map the prospect, plus the tail spread, plus the aperture window.
7) A 60-fold CMP program is planned to obtain the desired multiple attenuation and S/N enhancement. Two crews are available for the survey. Select a crew and calculate the survey cost. What will the shot-point spacing be to obtain 60-fold coverage?

Table 7.1. Crew for a 60 CMP program.

Crew	Energy Source	Streamer	Channels	$/km
A	Air Gun	2350 m	120	450
B	Explosive	3000 m	240	500

8) What (in words) is the result of making the maximum dip angle 5° compared with the value found in 1) above?

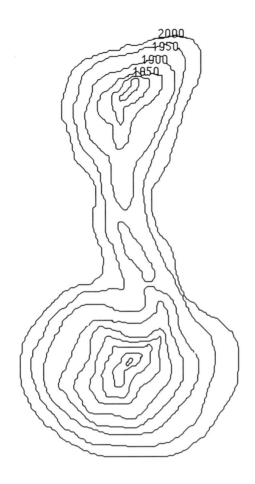

Fig. 182. A contour map.

Appendices

Appendix A

The Decibel Scale

The decibel (dB) scale is a method of comparing quantities, such as two signals, based on the logarithm of their power ratio.

By definition, the decibel level of signal #1 relative to that of signal #2 is given by

$$10\log\frac{P_1}{P_2},\tag{A.1}$$

where the Ps represent power. Since power is amplitude squared, the definition also can be given in terms of an amplitude ratio:

$$10\log\frac{A_1^2}{A_2^2} = 20\log\frac{A_1}{A_2}.\tag{A.2}$$

Two signals always differ by the same number of decibels whether they are described in terms of power or amplitude, because

$$10\log\frac{P_1}{P_2} = 20\log\frac{A_1}{A_2}.\tag{A.3}$$

Many parameters in geophysics vary over a wide range. For example, the voltage output from a geophone may be 100 000 times larger for a shallow reflection than for a deep reflection. Geophysicists prefer a logarithmic scale to a linear scale for calculations and display of such parameters. Also, we are usually more interested in the relative value of a parameter rather than the absolute value (how large it is at one time relative to another time).

Because of its logarithmic nature, the decibel scale provides a method to express conveniently large ranges of values. For example, a factor of 2 between two signal amplitudes is about 6 dB because $20 \log_{10} 2 = 20\,(0.301) \approx$ 6. Two signals that differ in amplitude by 2^{15} are 90 dB apart ($20 \log_{10} 2^{15} = 15(20)(0.301) \approx 90$). The following table shows some further equivalences between powers of 2 and decibel values.

Table A.1. Further equivalences between powers of 2 and decibel values.

4	=	2^2	=	2 x 6dB	=	12dB
8	=	2^3	=	3 x 6dB	=	18dB
16	=	2^4	=	4 x 6dB	=	24dB
32	=	2^5	=	30dB		
64	=	2^6	=	36dB		
128	=	2^7	=	42dB		
256	=	2^8	=	48dB		
512	=	2^9	=	54dB		
1024	=	2^{10}	=	60dB		
2048	=	2^{11}	=	66dB		
4096	=	2^{12}	=	72dB		
8182	=	2^{13}	=	78dB		
16 264	=	2^{14}	=	84dB		
32 728	=	2^{15}	=	90dB		
65 456	=	2^{16}	=	96dB		
130 912	=	2^{17}	=	102dB		

This table shows that when a value becomes large on a linear scale it is not so large using the dB scale. Thus, we can plot values of seismic signal and noise that differ by 90 dB more easily with the dB scale than with the linear scale. This scale is frequently used in many disciplines of geophysics and engineering.

To reiterate, if x and y are expressed in amplitude units, and $x = 2y$, then they differ by 6 dB. This is so because

$$20\log\frac{x}{y} = 20\log\frac{2y}{y} = 20\log 2 = 6. \qquad (A.4)$$

If power is computed, then

$$P_x = x^2 \text{ and } P_y = y^2.$$ (A.5)

The calculation gives 6 dB, since

$$10\log\frac{P_x}{P_y} = 10\log\frac{x^2}{y^2} = 10\log\frac{4y^2}{4y^2}.$$ (A.6)

Appendix B

Computing Array Responses

Consider a linear array of N receivers at constant spacing d where, by convention, the center of the array is represented by a flag.

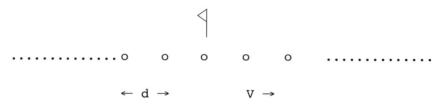

A wave with an in-line horizontal velocity V will arrive at each phone in turn at time intervals

$$\Delta t = \frac{d}{V} = \frac{d}{f\lambda} \qquad (\text{from } V = f\lambda). \qquad (B.1)$$

The Z-transform is a method used to represent a time series in either the time or frequency domain (Sheriff, 1991b). It is used now to solve the transfer function of an array.

The mathematical transfer function parameter for an odd number of receivers of an array is

$$
\begin{aligned}
A(Z) &= Z^{-(N-1)/2} + \dots + Z^{-1} + Z^0 + Z^{+1} + \dots + Z^{(N-1)/2} \\
&= Z^{(N-1)/2}[1 + Z^{-1} + Z^{-2} + \dots + Z^{-(N-1)}]
\end{aligned}
\qquad (B.2)
$$

where Z is the general progression transfer function parameter and the center receiver is at Z^0.

Note that the sum to N terms of a geometric progression is

$$S_N = a\left(\frac{1-r^N}{1-r}\right) \text{ for } a + ar + ar^2 + \dots . \qquad (B.3)$$

Let $a = 1$ and $r = Z^{-1}$. Therefore,

$$A(Z) = Z^{(N-1)/2} \cdot \frac{1-(Z^{-1})^N}{1-(Z^{-1})}. \qquad (B.4)$$

The frequency response is found from the Z transfer function by letting

$$Z = e^{i2\pi f \Delta t} \tag{B.5}$$

or

$$Z = e^{i2\pi d/\lambda}. \tag{B.6}$$

Hence,

$$A(Z) = e^{i\pi d(N-1)/\lambda} \cdot \frac{1 - e^{-i2\pi Nd/\lambda}}{1 - e^{-i2\pi d/\lambda}}. \tag{B.7}$$

Now,

$$1 - e^{-i2\theta} = e^{-i\theta}(e^{i\theta} - e^{-i\theta}) = e^{-i\theta} \, 2i\sin\theta. \tag{B.8}$$

Hence,

$$\begin{aligned} A(e^{i2\pi d/\lambda}) &= e^{i\pi d(N-1)/\lambda} \cdot \frac{e^{-i\pi Nd/\lambda} \cdot 2i\sin(\pi Nd/\lambda)}{e^{-i\pi d/\lambda} \cdot 2i\sin(\pi d/\lambda)} \\ &= \frac{\sin(\pi Nd/\lambda)}{\sin(\pi d/\lambda)} \end{aligned} \tag{B.9}$$

Thus,

$$\text{amplitude response} = |A(e^{i2\pi d/\lambda})| = \left| \frac{\sin(\pi Nd/\lambda)}{\sin(\pi d/\lambda)} \right|. \tag{B.10}$$

Because the amplitude value may vary widely, it is convenient to normalize peak amplitude values to unity by dividing by the number of phones. When normalized, we write

$$\left| \frac{1}{N} \cdot \frac{\sin(\pi Nd/\lambda)}{\sin(\pi d/\lambda)} \right|. \tag{B.11}$$

$$= Arg[A(e^{i2\pi d/\lambda})]$$

The phase response

$$= Arg\left[\frac{\sin(\pi Nd/\lambda)}{\sin(\pi d/\lambda)}\right] = 0 \qquad (B.12)$$

In this case the phase response is zero because the geophone spread is symmetrical. In the general case of a non-symmetrical spread, the phase will depend upon the geophone spacing.

Exercise B.1

Derive the general case for a linear array with an even number of elements.

Exercise B.2

Determine the expression for the phase response if Z^0 is the left side element, and discuss the effect on the phase response when $t = 0$ is moved to some other arbitrary reference point.

Appendix C

Weighted Arrays

When a weighted geophone array is used, the transfer function changes. For example, consider a weighted array with geophones of weighting 1,2,1:

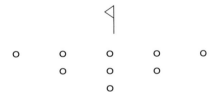

$$A(z) = Z^{-1} + 2Z^0 + Z^{+1}, \tag{C.1}$$

where the central 2 phones weight the array at the center. This is represented in the transfer function as a doubling of the central term of Z^0 to $2Z^0$.

The Fourier transform (FT) of these data takes them from the time domain into the frequency domain, where the transfer function is represented in terms of amplitude A and phase θ.

The frequency response is then:

$$A(e^{i2\pi d/\lambda}) = 2 + e^{+i2\pi d/\lambda} + e^{-i2\pi d/\lambda}. \tag{C.2}$$

Increasing the weighting further,

The expression becomes

$$A(z) = Z^{-2} + 2Z^{-1} + 3 + 2Z + Z^2. \tag{C.3}$$

For convenience, since wavenumber k is the reciprocal of wavelength λ, $z = e^{-i2\pi k \Delta x}$.

Then, the Fourier transform $= e^{+i2\pi k(2\Delta x)} + 2e^{+i2\pi k(\Delta x)} + 3 + 2e^{+i2\pi k(\Delta x)} + e^{-i2\pi k(2\Delta x)}$, where (Δx) and $(2\Delta x)$ are the offset distances to the receivers from the center.

Here is an example for irregularly spaced arrays:

```
              ⊲
              |

   o   6m   o    6m    2.5m    2.5m    6m   o   6m   o

   o        o      o       o       o       o       o

 -14.5    -8.5    -2.5     0      2.5     8.5     14.5
```

$$FT = 2e^{+i2\pi k(14.5)} + 2e^{+i2\pi k(8.5)} + e^{+i2\pi k(2.5)} + 1 + e^{-i2\pi k(2.5)}$$
$$+ 2e^{-i2\pi k(8.5)} + 2e^{-i2\pi k(14.5)}, \tag{C.4}$$

where

$$\theta j = 2\pi k x i \text{ when } x \text{ is a negative distance from the take-out,}$$
$$= 2\cos\theta_j + 2i \sin\theta_j + \ldots + 2 \cos\theta_j - 2i \sin\theta_j. \tag{C.5}$$

All imaginary parts cancel, leaving real cosine terms only.

Unequally spaced phones are easily accommodated by modifying Δt, so that all times are multiples of a common unit.

An example of where the spacing of each term in the array is $d/2$:

```
              ⊲
              |

        3d           d         3d
   o    .    .    o    o    .    .    o
```

$$A(z) = Z^{-7} + 0 + 0 + 0 + 0 + 0 + Z^{-1} + 0 + Z^{+1} + 0 + 0 + 0 + 0 + 0 + Z^{+7} \tag{C.6}$$

where

$$\Delta t = \frac{d}{2V} \tag{C.7}$$

and

$$f\Delta t = \frac{d}{2\lambda}. \tag{C.8}$$

Appendix D

Fourier Analysis

Fourier (1786-1830) theory states that a nonperiodic time function (e.g., a seismic trace) may be treated as the sum of a series of harmonic waves, each having a specific amplitude, frequency, and phase. The Fourier transform is a mathematical procedure that decomposes a signal into its harmonics. The harmonic representation of a signal is called its frequency-domain representation. The inverse Fourier transform applied to the frequency-domain representation of a signal, such as a seismic trace, converts the signal into the time domain. The three parameters which uniquely characterize a harmonic wave are:

Amplitude—The maximum excursion from a zero value. In Figure D-1, the amplitude of waves A, B, and C are 10, 10, and 5 units, respectively. For a seismic trace, or any other signal, amplitude may be expressed in terms of voltage.

Frequency—The number of times per second a wave repeats itself, which is the same as the number of positive amplitude peaks that occur per second. This may be expressed as the number of cycles per second or as *hertz* (Hz). We also call the time from one positive peak to the next (or from one negative peak to the next) the *periodic time*.

$$F = \frac{1}{T}, \tag{D.1}$$

where F is the frequency and T is the periodic time.

Since wave A in Figure D-1 displays two cycles in 0.1 s, then its frequency is 20 cycles per second or 20 Hz. Wave A's period is therefore 1/20 s. The frequencies of waves B and C are both 10 Hz, with periods of 1/10 s.

Phase—The difference expressed in angular measure between a reference point in time and a chosen point on a wave. For example, in Figure D-1, a chosen point on a peak on wave C must move 0.025 s to the left to make the point appear at zero time. That is, if zero time is the reference point for peaks on wave C, then wave C lags the zero reference point by 0.025 s. This is known as having *negative phase*. One full cycle of a wave corresponds to a phase of 360°. Since the period of wave C is 0.1 s, its phase lag is (0.025/0.1) x 360°, or 90°.

The Fourier transforms in Figure D-2 show:

1) An infinite range of frequencies will be produced from a spike in the time domain.

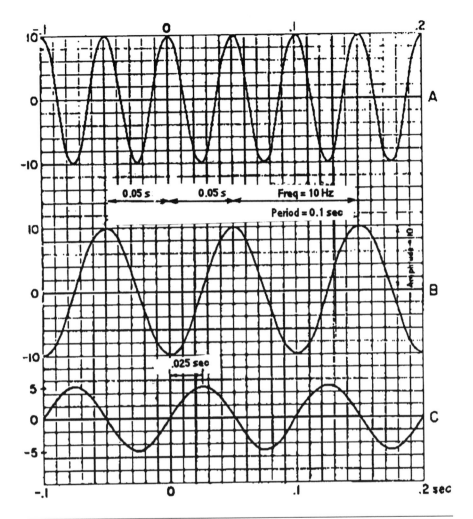

Figure	Amplitude	Frequency	Phase-Radians	Phase-Degrees
A	10	20 Hz	0	0
B	10	10 Hz	$\pm\dfrac{0.05}{1/10} \times 2\pi = \pm\pi$	$\pm\dfrac{0.05}{1/10} \times 360 = \pm180$
C	5	10 Hz	$-\dfrac{0.025}{1/10} \times 2\pi = -0.5\pi$	$-\dfrac{0.025}{1/10} \times 360 = -90$

Fig. D-1. Amplitude, frequency, and phase.

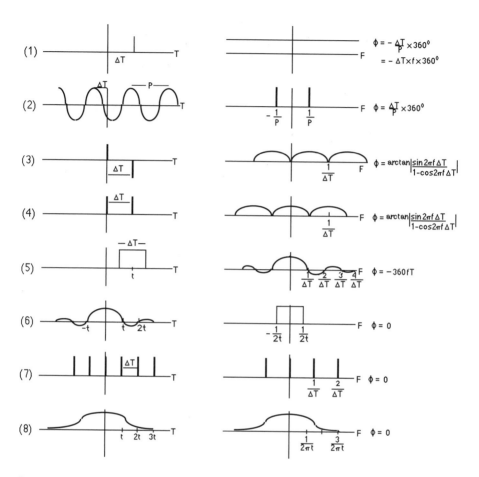

Fig. D-2. Fourier transforms.

2) A single frequency in the frequency domain is the result of an infinitely long wave in the time domain.

3) An impulse of two opposite sign spikes in the time domain is a filter with notches at $1/\Delta t$.

4) Two positive spikes in the time domain produce notches at $1/2\Delta t$, $3/2\Delta t$, and so on in the frequency domain.

5) A block time function has low frequency with a central peak at DC and side lobes that decay. A closely spaced geophone array is a filter with this shape.

6) A block frequency-domain filter has a (sin x/x) time-domain impulse response.

7) A comb wave shape in time is a comb shape in frequency. Shorten tooth spacing in one domain and it spreads in the other.

8) A bell-shaped wave which tapers to zero at each side of the center of the bell in the time domain produces a similar frequency-domain shape.

Phase Response and Types of Phase

The Fourier transform decomposes a wave into its phase and amplitude content at different frequencies. A phase spectrum can be displayed graphically by plotting phase versus frequency. Often, for convenience, a wraparound display is used in which 360° is subtracted from the true phase any time it exceeds 180° (see Figure D-3).

Seismic wavelets can have different types of phase. We can separate wavelets into five categories of phase, which are called linear-, minimum-, maximum-, mixed-, and zero-phase. These are described in Figures D-4 through D-8 as follows:

The phase spectrum in a seismic wavelet depends on:

1) Source character
2) Receiver response
3) Earth's characteristics
4) Recording instrument impulse response
5) Data processing parameters

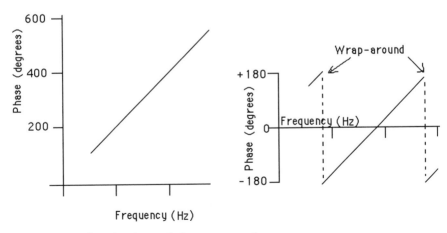

Fig. D-3. Graphic displays of phase versus frequency.

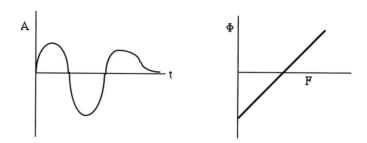

Fig. D-4. Linear-phase—The phase of each harmonic is defined by a constant times the harmonic frequency. The phase spectrum is a straight line passing through the phase origin.

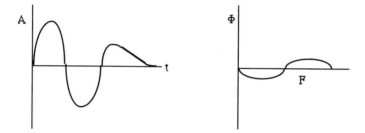

Fig. D-5. Minimum-phase—A function which concentrates energy in the time domain close to zero without having energy present before time zero. The phase value is low at all frequencies.

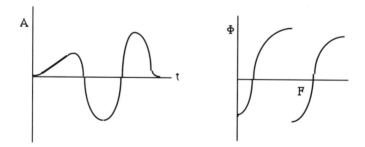

Fig. D-6. Maximum-phase—Energy concentrated as far from time zero as possible (that is, at the end of the wavelet) produces a maximum value of phase at all frequencies (shown here as wrapping around).

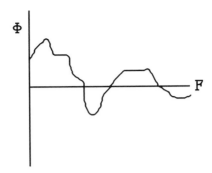

Fig. D-7. Mixed-phase—Any other function of phase not covered above.

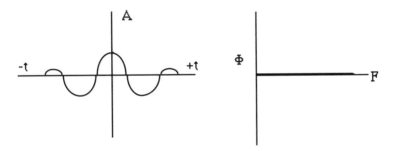

Fig. D-8. Zero-phase wavelet.

The dynamite source is minimum-phase since all energy is exploded on detonation (this is sometimes referred to as being front-end loaded in energy). Air guns are very close to minimum-phase for the same reason but have a slightly broader energy spread in time and frequency (see Chapter 3). Vibroseis signals are zero-phase after crosscorrelation with the sweep signal. The source ghost (the reflection of the air-Earth interface) which travels down and behind the source wavelet tends to minimum-phase while energy absorption by the Earth causes the phase to lose its minimum-phase value.

When the source, receiver, and instrument phase relationships are known, we can alter their character by computer processing to help our interpretation. Ideally, the interpreter needs a wavelet centered at the horizon of interest to him. For convenience, a wavelet having no phase (i.e., zero-phase) response is used, and so we convert all reflected events to a zero-phase wavelet wherever possible. A zero-phase wavelet is centered at zero time and symmetric, as shown in Figure D-8.

Exercise D-1

For the following waves, compute the frequencies and note how, when they are summed to produce the wavelets at the bottom of the two figures, the wavelet shape changes.

At the point where the waves coincide to become in-phase (as in Figure D-9), they may sum coherently to produce a short-period wave. Figure D-10 shows that many more frequencies are required to be summed together to produce a narrower wavelet.

These figures help explain how bandwidth affects resolution. A narrow-frequency wavelet would not detect thin beds or small faults, whereas a broad-band wavelet consisting of many frequencies (as in Figure D-10) may carry information of subtle geologic (facies) changes.

The received signal may have a poor quality waveform, but the recording instruments must reproduce this faithfully. Any reduction in recordable bandwidth would reduce the useful frequency content available for data processing. The greater the bandwidth of frequencies which have been put into the ground, the greater the chances of recording broad-band reflection energy. Hence, the energy source should produce a broad-band source signal.

TIME (seconds)

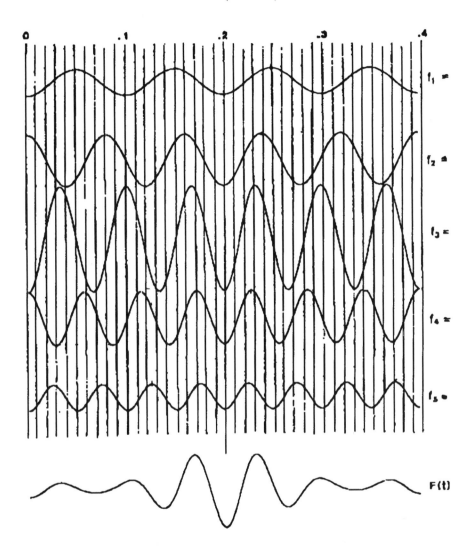

Fig. D-9. A short-period wavelet is produced at a point where waves are in-phase.

TIME (seconds)

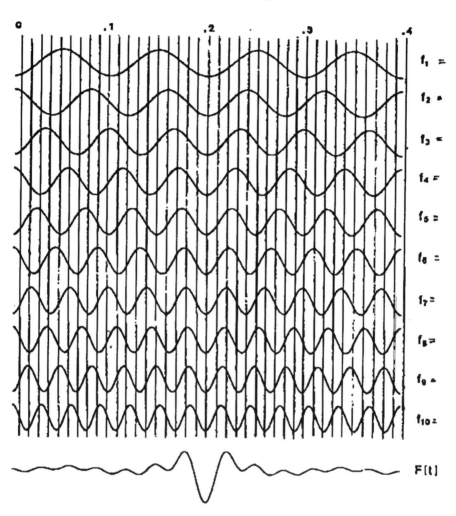

Fig. D-10. More frequencies are required to produce a narrower wavelet.

References

Aki, K., and Richards, P. G., 1980, Quantitative seismology—Theory and methods, Vol. 1: Freeman.

Allen, J., and Howell, J., 1987, Using "Poor man's 3-D" to identify distributary channel sands in the Wilcox formation, Lavaca County, Texas: The Leading Edge, **6**, No. 10, 8-15.

Anstey, N. A., 1986a, Whatever happened to ground roll?: The Leading Edge, **5**, No. 3, 40-46.

Anstey, N. A., 1986b, A reply by Nigel Anstey: The Leading Edge, **5**, No. 12, 19-21.

Arnold, M. E., 1977, Beam forming with vibrator surveys: Geophysics, **42**, 1321-1338.

Aston, M. W., 1977, Theory and practice of geophone calibration in-situ using a modified step method: IEEE Transactions on Geosciences and Remote Sensing, **15**, No. 4, 208-214.

Bullen, K. E., and Bolt, B. A., 1987, An introduction to the theory of seismology: Cambridge Univ. Press.

Barr, F. J., Wright, R. M., Sanders, J. I., and Obkirchner, S. E., 1990, A dual-sensor, bottom-cable 3-D survey in the Gulf of Mexico: 60th Ann. Internat. Mtg., Soc. Expl. Geophys. Expanded Abstracts, 855-858.

Cerverny, V., Molotkov, I. A., and Psencik, I., 1977, Ray method in seismology: Univerzita Karlova.

Cunningham, A. B., 1979, Some alternate vibrator signals: Geophysics, **44**, 1901-1921.

Dobrin, M. B., Simon, R. F., and Lawrence, P. L., 1951, Rayleigh waves from small explosions: Trans. Am. Geophys. Union, **32**, 822-832.

Dragoset, W. H., 1990, Air-gun array specs: A tutorial: The Leading Edge, **9**, No. 1, 24-32.

Evans, B. J., Oke, B., Urosevic, M., and Chakraborty, K., 1995, A comparison of physical model with field data over the Oliver field, Vulcan Graben: APEA Journal, **24**, 26-43.

Evans, B. J., and Uren, N. F., 1985, The seismic performance of shaped charges: Minerals and Energy Research Institute of Western Australia, Report on Project 66.

French, W. S., 1974, Two-dimensional and three-dimensional migration of model-experiment reflection profiles: Geophysics, **39**, 265-277.

Goodfellow, K., 1990, Special Report: Geophysical Activity in 1990: The Leading Edge, **9**, No. 11, 49-72.

Goodfellow, K., 1991, Special Report: Geophysical Activity in 1990: The Leading Edge, **10**, No. 11, 45-72.

Hydrographer of the Navy, 1965, Admiralty manual of hydrographic surveying, Vol. 1.

Kennett, B. L. N., 1983, Seismic wave propagation in stratified media: Cambridge Univ. Press.

Kramer, F. S., Peterson, R. A., and Walter, W. C., Eds., 1968, Seismic energy sources 1968 handbook: Bendix United Geophysical Corp.

Lansley, M., 1992, Personal communication.

Lynn, W., and Larner, K., 1989, Effectiveness of wide marine seismic source arrays: Geophys. Prosp., **37**, 181-207.

Martin, J. E., and Jack, I. G., 1990, The behavior of a seismic vibrator using different phase control methods and drive levels: First Break, **8**, 404-414.

Mayne, W. H., 1962, Common reflection point horizontal data stacking techniques: Geophysics, **27**, 927-938.

McCready, H. J., 1940, Shot-hole characteristics in reflection seismology: Geophysics, **5**, 373-381.

McDonald, J. A., Gardner, G. H. F., and Kotcher, J. S., 1981, Areal seismic methods for determining the extent of acoustic discontinuities: Geophysics, **46**, 2-16.

Nestvold, E. O., 1992, 3-D seismic: Is the promise fulfilled?: The Leading Edge, **11**, No. 6, 12-21.

Pilant, W. L., 1979, Elastic waves in the earth: Elsevier.

Rietsch, E., 1981, Reduction of harmonic distortion in vibratory source records: Geophys. Prosp., **2**, 178-188.

Sheriff, R. E., 1991a, Arrays, Fig. A-15, Encyclopedic dictionary of exploration geophysics: Soc. of Expl. Geophys.

Sheriff, R. E., 1991b, Encyclopedic dictionary of exploration geophysics: Soc. of Expl. Geophys.

Society of Exploration Geophysicists Technical Standards Committee, 1980, Digital tape standards: Soc. of Expl. Geophys.

Stansdell, N. A, 1978, The transit satellite navigation system: Magnavox publication, California.

Steeples, D, 1992, High resolution seismic surveying short course notes: Soc. of Expl. Geophys.

Stone, D. G., 1994, Designing seismic surveys in two and three dimensions: Soc. of Expl. Geophys.

Vermeer, G. J. O., 1991, Symmetric sampling: The Leading Edge, **10**, No. 11, 21-29.

Yilmaz, O., 1987, Seismic data processing: Soc. of Expl. Geophys.

Index